12/19

Now That Was A Day To Remember!

To
Joan
Happy Trails
Dan Cossa

DAVID COSSA

First Edition

PAGE PUBLISHING, INC.
New York, NY

First originally published by Page Publishing, Inc. 2018

ISBN 978-1-64298-525-2 (Paperback)
ISBN 978-1-64298-526-9 (Digital)

Printed in the United States of America

Contents

INTRODUCTION

So why Mountain Dave? Once I was hiking the Northern Cascades section of the Pacific Crest Trail and made camp near a couple that was through hiking the entire trail. There was another person named David in the camp, so they differentiated me by calling me Mountain Dave when they found out I climbed mountains. Everyone that through hikes the Pacific Crest Trail gets a "trail name," and that is the one that stuck.

I owe all my outdoor adventures to my mother and grandmother, who insisted that we move to her home State of Colorado from the East Coast when I was six. We lived right on the very edge of Suburbia, just farmland to our north and west. That meant as a kid I was able to follow the railroad tracks near our home in summer and dry irrigation ditches in winter for many miles to see where they went, stopping to swim in farmer's ponds along the way. Once three of us, ages around twelve, hiked the entire way to where the Rockies jut up from the High Plains, some fifteen miles away and most of the way back before our parents picked us up. Bike rides into the Rockies followed where we camped along Coal Creek. Then came the big one—the fifty-mile hike challenge offered by President Kennedy as part of a physical fitness for youth campaign. You got a medal if you could hike the distance in twenty-four hours or less. A friend of mine and I walked all the way to Boulder, some twenty-five miles away, and about twelve miles back before I made the fatal mistake of sitting down. My legs didn't cramp, they simply froze up, and I couldn't get up. It was like they were paralyzed. My friend went to a nearby house, they called my parents, and when they came out they had to pick me up by my arms and place me in the car. My legs were fine the next morning. My friend did finish the hike but barely in time and got his medal. I was very disappointed that I didn't make it.

One could say that was the beginning of my interest in long-distance hiking.

To my grandmother, I owe two other things—my roots for mountaineering because she insisted I be sent to her church's camp high up in the Rockies. One of their outings was an ascent of nearby Mount Audubon, just a walk up, but quite an introduction to the alpine world above timberline, to be able to slide and walk over a permanent snowfield in August, the view of the Never Summer Range mesmerizing and dominated by nearby Longs Peak—that seemed to beckon come even higher!

So I did several years later climbing to a place known as the keyhole on Long's Peak with three high school buddies. We went no further because of snow covering the slopes that led to the summit, and that made me realize that if I wanted to advance in this sport there was more to it than just going—we had no ice axes, and that was that.

The other thing I owe my grandmother for is the guardian angels she no doubt prayed would protect me—and did they ever, as you shall see.

THE NAVY

Flight Deck of USS John F Kennedy

My first close call took place while I was on an aircraft carrier, the John F. Kennedy, in the Mediterranean Sea. I was an aircraft electrician in a radar plane squadron. Now it must be said that for the most part, things were fairly dull in this particular squadron. Something was always wrong with the sensitive radar electronics. The squadron was shorthanded when it came to electronic technicians. The only one they had was an Okie from Muskogee type. He would just throw his hands up and say in a Southern drawl "I can't fix that problem sir. They never sent me to that electronic school that deals with that there problem."

The only reason to fly our planes was for their electronic surveillance capability. They were sort of an airborne traffic control system used to identify all aircraft in a given area. Since the electronics didn't work, the planes didn't waste fuel flying around for no reason. Since they never flew around much, nothing much went wrong with them electrically. Since there were no electrical problems, there was no work for the electricians including me. We spent most of our sixteen hour days watching movies, playing cards, listening to our stereos, or roaming about the giant ship. Perhaps all this inaction caused us to be less than safe. Occasionally, the planes did fly to keep our pilots qualified at whatever they were supposed to be qualified at.

One time a plane returned with a landing gear light problem. A warning light in the cockpit told the pilot whenever the landing gear was in transition from down to up and vice versa. When down and locked in place, the light went out. Similarly, when the gear was up and locked in place the light went out. The pilots reported that the light was on when the gear was supposedly down and locked. Since doubt was cast on this condition via the warning light, an emergency landing was performed. The landing gear did not collapse upon landing, so obviously there was a problem with the electrical part of the warning light system.

The day shift electricians checked out the micro switch on the landing gear. The switch was supposed to be in the off position when the landing gear was down. They found that was indeed the case and turned the problem over to the night crew, meaning me. I was to check for wiring problems in the cockpit lever that caused the landing gear to go up and down. This could only be done with both propeller engines running. There were some minor mechanical problems as well so three machinists went into the cockpit with me and one of them started the engines. A fourth mechanic climbed up into the nose wheel well. A plane captain stood guard with a fire bottle in case of engine trouble.

After the engines started, I looked at the warning light, and sure enough, it was on. The manual stated that the lever was to be pulled as if a pilot was raising the gear and observe what happened to the light. The landing gear wouldn't actually rise because braces were placed on them by the plane assistants while the plane was parked to prevent such an occurrence.

I grabbed the lever and started to engage it. As I did so, the thought entered my mind, I wonder if those braces are actually. The plane lurched downward. The plane captain had forgotten to put the braces on, and apparently, the day shift electricians who were working on the landing gear had failed to notice this oversight. I immediately jammed the lever back into its original position, and the plane rose back up again after having fallen a couple of feet. The plane captain said the propellers had come within inches of the flight deck. If they had touched the flight deck, the engines would have exploded and most likely everyone on or near the plane would have been killed by shrapnel or the fireballs that would have ensued. The mechanic in the nose wheel said that his arm was just beginning to get pinched by the rising nose gear before it returned to the down position. The mechanics immediately shut down the engines, and we all left the plane without saying a word. We were all badly shaken by the near catastrophe.

Much to my surprise, an inquiry into the incident found only fault with the plane assistant who was supposed to routinely put braces on the landing gear while the plane was parked—it was standard procedure. He was fined two weeks' pay. Since the electricians had nothing to do with these braces, no fault was taken with us, but they amended the manual after this incident to require us to visually check for the braces before following the trouble shooting procedure with the landing gear lever.

The flight deck of an aircraft carrier is the most dangerous place on Earth. The jets will suck you in or blow you overboard, a two-inch thick arresting cable is no fun when it snaps under high pressure, sometimes the catapult just doesn't work right or you are overloaded and into the ocean you go only to be run over by the ship. Landing plane's wing tips are just a few feet from parked planes and once in a great while they miscalculated and hit them. They land one a minute during drills, you have sixty seconds to get the plane out of the way after landing.

Because we were idle so much of the cruise, once in a while, I would go sit in a parked plane's cockpit because I had legal access to them, and watch them land. Not once could I not turn my head away as they went by. They looked like they were going to land right on top of me every last time.

In fact, they were sort of landing right on top of me—when I slept. My top bunk on the 03 level was only about eighteen inches below the bottom of the flight deck, right on the fantail where they landed. If the deck had been made of Plexiglas I could have watched them bounce down right on top of me. It was hard to sleep during exercises, every minute there would be a big BANG! followed by the sound of the arresting cable slithering back into place to await the next plane.

We lost five planes on our cruise about normal from what I understand. Two flew into a mountain range in the fog, two went into the ocean after launching and one broke its tie down chains while testing the engine under full power and over the side it went. Overloaded mail plane that went down while launching was worst four people on it, and it got run over by the ship with the expected results. The pilots in the other plane that hit the water ejected and were rescued with no injuries. They shared the same Ready Room as our squadron did and had quite a story to tell as we watched the video of their plane crashing and their ejecting, over and over again.

When in port, members of the Air Wing were drafted to stand shore patrol, and when in Athens and Barcelona I was one of them.

Nothing memorable happened in Barcelona except that it was one hell of a beautiful city. Athens however was quite memorable, and not just because of the Parthenon tour where I tagged along just in earshot of an English-speaking guide, learning all kinds of things about the place for free.

Some of the memories were of just how drunk sailors could get while in port to let off steam. For example, one night I was standing shore patrol where sailors came back to board the launch to go back tom the carrier. The launches came and left once an hour. Part of my job was to make sure no alcohol is brought back to the ship by checking packages if I felt the need to, confiscating any liquor found and issuing a citation.

A chief who was obviously inebriated given the way he walked toward me, had a sort of plan to get around that.

"Shore Patrol, eh?" he blurted out. "Well watch this!"

He took a fifth of Jim Beam out of a paper bag and lifted up the bell bottom of one of his pant legs. He then stretched out his sock *real* wide and stuffed the whiskey bottle in it. Then over the whole thing the bell bottom went, a huge bulge still visible. He stumbled over to the quarterdeck, saluted, and asked the quartermaster for permission to board ship. The quartermaster, who was doing this all the time, never noticed the bulge. Then, as he walked down the gangway to the launch, the chief stopped, looked at me, and gave me a big wave. I just shook my head and laughed at his nerve—and let him go. He somehow knew I would. I had heard that most chiefs had their private illegal liquor stashes aboard, and it was overlooked as a perk as long as they didn't drink during working hours. It wasn't working hours so I overlooked it.

And who could forget the sailor coming back from the wine tasting festival. I had been there the night before. For a very reasonable fee, they gave you a Greek wine glass and said have at it! Casks of wine were in a long row with spigots on them. At one end was the driest wine imaginable, it sucked the saliva right out of your mouth, and on the other end the wine was so sweet it was almost like a grape syrup.

Well this guy had gotten his money's worth for sure! He was too drunk to walk, so he was being carried by two other sailors, one under each of his arms. You could tell it was wine he was drinking because first he would spew a perfect arc of purple vomit one direction, then turn his head and spew out another perfect arc, like some sort of purple fountain.

Okay, this might be grossing you out; so let's get back to the Parthenon. Did you know they hollowed out the columns and poured molten lead down the holes to act as a sort of rebar when it hardened? I didn't either until the free tour.

Getting back to the Navy, the last night in port I was assigned to stand shore patrol at some obscure strip joint in Athens. After being dropped off, I was told to make sure to catch the last launch back to the ship at 2:00 a.m., because we were leaving port the next morning. I was about a mile inland from where the launch vessels tied up to the pier. The bouncer there was a tall lean guy named Frank.

FRANK THE BOUNCER

Frank was the only guy around who knew English, who said he was from the US. Very few sailors showed up at this rather obscure place. There was no way to know about it unless you stumbled upon it while bar hopping. So Frank was my main company for the night and he was very outgoing and talkative.

We talked all evening and it turned out Frank was a likable, nice guy. The local drunk came down the street around midnight so soused he simply walked in a straight line across the road until he hit something and fell down. Blind drunk.

He repeated the process three times hitting his head fairly hard on the sidewalk the last time he fell down, when Frank said that's enough and hailed a cab for the poor guy to take him home. Frank paid the fare saying he'd pay me back tomorrow when he's sober. Around 1:00 a.m. it was time to go to make certain I caught the last launch out of Athens for the cruise—I had to walk the mile back. I said bye Frank, and as I walked down the street, I turned around to wave goodbye.

"Hey, kid!" he yelled. "When you get back to the ship, look up June 11, 1962, in the almanac!"

"Why?" I yelled back.

"You'll find out why," he said.

The ship's library was closed the next day, a Sunday, and we left port headed back to the US after a nine-month cruise. On Monday I went to the library after work, got the almanac for 1962, and looked up the date.

An escape from Alcatraz was the only entry. Three men—one named Frank Morris—had gotten out and were presumed drowned in the Bay—bodies never found. Well, I knew where Frank ended up!

I figured it must be true because why would someone say something like that out of the blue just for the hell of it. No, I never bothered to try and turn him in. I had no idea who to contact being on a ship in the middle of the ocean, and they likely would not have believed me anyway—not to mention the language barrier. I thought to myself what a perfect hideout—working as a bouncer in Athens of all places.

MEET THE WITCHES

The following stories are why I have come to believe in witches. I consider myself to be a very logical person. Before these incidents, I scoffed at the thought of the supernatural being anything but the figment of someone's imagination. Now, at least, I'm not so sure.

One Friday morning, while in the Navy stationed in Norfolk, Virginia, I was told by a superior officer that I would be flying out to an aircraft carrier for an extended two-month tour of duty. Although someone else in my squadron was more qualified to participate in the task at hand, he was on leave for the next two weeks. They needed someone now, and I was it.

"Go pack your bags right now," I was ordered. "Your flight leaves in four hours."

The wheels in my brain started turning as I struggled to find a way out of this undesirable situation. "Um, um, um, but, sir!" I stammered. "I can't go this afternoon. I just bought a new car and haven't bought any insurance for it. I can't just leave it out in the parking lot. You know how enlisted men are. They get drunk all the time and one of them is likely to plow into my uninsured car when they return from a binge at some bar. However, I do have an uncle who lives about two hours away. I could drop my car off at his house, but I wouldn't be back in time to make that afternoon flight. Would there be any possibility of catching a later flight?"

The officer looked at me and said straight up, "You're lying!"

"No, sir," I insisted. "See? Look out the window. That yellow Volkswagen out there is the car I'm talking about."

"He's right, sir," someone chimed in. "He just bought it a few days ago."

The officer considered the situation. Perhaps to show that he was capable of some compassion, he finally said, "Okay, there's another flight leaving on Sunday at noon. Make sure you're on it."

The officer was right. I was lying. The Volkswagen was fully insured. Instead of driving to my Uncle's house, I immediately drove up to Washington, DC, with a friend from there, intent on partying before putting in my time at sea. That Saturday night we visited a crowded bar in Georgetown. As we sat there boozing it up, two attractive young women came in, but all the tables were taken. They asked if they could sit at our table, since we had two empty seats.

Needless to say we said, "Sure! No problem!"

After talking to them for a while, one of them blurted out, "We're witches."

They didn't have the appearance of one's traditional view of witches, and I said so. "We're good witches," they said, "not wicked ones."

Yeah right, I sneered in total disbelief. "And this is the land of Oz! Right? Okay, if you're witches, how about doing me a favor by fixing it up so I don't have to go out to sea tomorrow." One of them looked me straight in the eye, smiled and said with utter confidence, "Consider it done! You were kind to us and now we will be kind to you!" Just like that. "Consider it done!"

After meeting our objective of partying until the bar closed, I arrived back at my barracks in Norfolk at 4:00 a.m. to find a note on my room door informing me of my flight time. It had been moved up from noon to 8:00 a.m. After only two hours sleep, I sadly packed up and traipsed on over to the air base airport and checked in. "The flight has been delayed for a while," I was told. "Mechanics are replacing a part. Go sit down in the waiting room and we'll give you a call when it's time to board the flight."

I did as I was told. Since I was up all night and had a hangover too, I quickly fell asleep. When I awoke I glanced at my watch. Several hours had passed. I went back to the boarding area to see what was cooking. "How're those repairs going," I asked. "Holy cow," they said. "We forgot all about you! That flight just left and there isn't another flight leaving for the carrier for [flip, flip, flip, flip] two weeks."

"Man," I said, "I'm going to be in big trouble for missing that flight."

"It wasn't your fault. We said we'd come get you. If you like we can right up a paper for you explaining the circumstances."

"I would sure appreciate that," I said. I took that paper back to the same officer who had ordered me out to sea. He crumpled it up in disgust and threw it in the wastebasket. He had no reason to discipline me since I had done exactly as I was told to do. Since the more qualified man was back from leave before the next flight out to the carrier, he was sent instead.

Coincidence, you say? Perhaps, but another incident convinced me that witches do indeed exist. While at a party many years later, I met another self-proclaimed witch. She was a bit tipsy and "slipped" to reveal her powers. "Let's play a game," she said. "I can guess your birthday."

Before I could say anything, I had the bizarre feeling that someone was rummaging through my brain like a clerk going through a file cabinet, and she blurted it out. Then she guessed the birthday of another person listening to this going on and got the year right, but missed it by one day. Since I'm so logical, I tried and tried to figure out how she pulled off this parlor trick, but there is simply no logical explanation of how she did what she did. There was no way she could have known my birthday. No one at that party knew it either. I know it sounds ridiculous, but the only possible explanation was that she read my mind. How? Who knows?

THE COMMUNE

After I got out of the Navy, I attended Community College of Denver under the GI Bill. I met another vet there who said they had room for another person in the "commune" he was living in. In reality it was eight people living together in the same house, sort of a cross between Doonesbury's Walden Pond and Friends.

It was a large old five-bedroom two-story Victorian home, so common in Denver. There was one slight hitch though. The property was in the heart of Denver's ghetto, Five Points. For some reason, the place was owned by a church and every month someone in a suit would show up for the $200 rent—split eight ways. Life was good I had a job delivering the Rocky Mountain News on Capitol Hill to a bunch of apartment buildings. My whole route was only about five square blocks so it was easy spending money. I finished about seven each morning, spent an hour at a restaurant having breakfast and then off to school! All of us had jobs somewhere and things were good for everyone with such low rent.

To say where we lived was a high crime area was putting it mildly, but no violence toward us, just a couple of burglaries and we quickly learned what streets not to drive down lest the kids throw rocks at you for being White. We heard gunshots every weekend it seemed, family arguments, robberies of business, fights at bars, that all turned deadly. A man even died under one of our cars after he got shot by the cops at a bar, ran away, and hid.

One time I came back after school with another house mate, and when we walked in the door we noticed cardboard boxes with all kinds of things in them like small appliances. We realized we had just interrupted a burglary in progress. We ran to the back door, and I saw a kid about eight years old crawling down the garage downspout.

He had jumped onto the garage roof from a second story window of the house—an easy jump—in an attempt to escape. I ran over, jumped up, grabbed him, and pulled him down.

I said, "You've been here before, haven't you!"

I had a large jar of pennies and other items that I noticed were missing about a week prior. He wouldn't say a word. I asked where he lived, still nothing. So I had an idea.

I said to him, "You see that chain-link fence there? If you don't start talking I am going to TOSS you right THROUGH it at the count of three. You'll end up being sausage! One! Two!"

His eyes got as big as saucers because he thought I meant it, and he started up with a nonstop babble—places, times, his name, where he lived, so my bluff worked. He lived about two blocks away, but I had to drag him right through the heart of the ghetto business district, getting all kinds of perplexed looks as to what the hell I was doing a white guy with this black kid by the arm When I got to his house, I knocked on the door and his mother answered it.

When I said, "I just caught your kid burglarizing my house!" she simply said, "I can't do nothing with him!" and slammed the door. Still angry, I called the cops.

A seasoned thirty-year detective came out, and I said, "I want to press charges!"

The cop looked at me and said, "Are you SURE you want to do that?"

"Why not?" I said.

"Because this kid has been to court eight times, and nothing has stuck yet. You are free to do that, but remember one thing, he's got friends who will get even for him. What do you want me to charge him with again?"

Needless to say I said, "Let's just forget the whole thing."

The commune finally fell apart when one of the women decided to get religion and joined some Hindu Guru cult that was having a "seminar" in Denver—the twenty-one-year-old perfect guru or something like that.

"God is here on Earth, isn't it cool!" she bubbled. "Come on, let's go see him!"

So I let her drag me along to a house teeming with his followers and the dude wouldn't even come out of his bedroom when they all chanted at his door how much they loved him and to please come out and bless them—too busy being a "playboy" probably because his mom kicked him out of the cult for being just that a few months later.

The convert made the mistake of inviting a few over who said they needed a place to stay for the night and before you knew it thirty of them had crashed the place. We of course ask them to leave, but they said they were "invited" to stay and had no intention of leaving. Vastly outnumbered there wasn't much else we could do but move out or put up with hour long waits at the way overused bathroom. We couldn't throw them out without going to court and that would have taken months.

About that time the following happened.

ART THE FART

Old Faithful Inn

Okay sorry about the obscenities in the following two stories, I just thought they were necessary to portray a sense of the true atmosphere that existed, so fair warning.

Hey, David, they're hiring workers for Yellowstone. All you gotta do is go is go down to the National Park Service office at the Denver Federal Center and they'll hire you on the spot! I was just about to go on summer break from the community college I was attending and was seeking summer employment. So I did indeed go to the NPS office and they indeed did hire me on the spot. I was a veteran and that was the only criteria for the job. I myself and two other vets, Larry and Bill, were given one week to show up in Yellowstone for a surveyor's aide job. Having absolutely no experience in this field, we had no idea what to expect, but a summer job in Yellowstone romantically appealed to me.

After packing up and notifying our landlords, we headed out in my 1970 Volkswagen, for parts unknown. It took only one day to drive there and check into the engineering office. Now it should be noted that this was in 1972 when discrimination against long hair was rampant and we indeed looked the hippie type. After having been forcefully groomed for years in the Navy we all were greatly enjoying letting our freak flag fly. The rest of the world, however, didn't seem to appreciate our new found freedom including our red neck co-construction workers. To say they were dismayed at our appearance is putting it mildly. A heavy equipment operator immediately shaved his head in protest. We were given berthing in a trailer park and our neighbors informed their kids not to accept any candy from us lest it be laced with LSD.

The head engineer was in a dilemma. There are several parts to survey team; an engineer, an instrument man and chain men or surveyors' aides. He had no instrument men whatsoever. We soon learned that these remote NPS jobs were difficult to fill due to the severe isolation. Very few family men were willing to separate from their loved ones for the lousy pay offered by the NPS. Family housing was very scarce. Our team engineer was sort of a liaison between the head engineer and the rest of the team, and he confided to us that the head engineer was distraught because he desperately needed instrument men and all he got was three inexperienced hippies from Denver.

The situation was this: we were to relocate several stretches of highway each several miles long. They had plenty of heavy equipment and operators for that equipment, but everyone was idle because the surveyors had yet to tell them where to put in the new road bed.

So during our first week there not much was accomplished. We visited the local store and restaurant which were stocked with dozens of college aged women who liked to flirt with the new hippies. Other park employees apparently became jealous. They complained to their supervisors as to why they were not allowed to have long hair and we were. This all got back to the head engineer and he called us in one morning and told us we had to get haircuts or we were to be fired (which was what he really wanted to do anyway). But we had an ace up our sleeve. There was a new clause in our hiring contract that stated there was to be no discrimination due to race, creed, color, ethnic origin or any other nonmerit factor. Non merit factor basically meant long hair. Bill gleefully pointed this out to him and the head engineer felt he had no option but to relent. His retreat, of course, did not endear him any closer to us.

But he had a secret weapon up HIS sleeve, too, in the form of one nasty, red necked, opinionated, loudmouthed, mo——fo—— cowboy that we soon nicknamed "Art the Fart," Mr. Fart got along with NOBODY, much less three long-haired hippies. He was brought in special to teach us to be instrument men in the hope that working with him would be so unpleasant that we would all go running back to Denver with our hair tucked between our legs. We sparred using politics right off the bat. 1972 was an election year and Nixon was running against McGovern. I'm guessing you know who supported who. One of our first conversations went something like this: Art the Fart—"So you lowlife motherfucking hippies are gonna vote for that good-for-nothing, yellow-bellied, piss-ass, bleeding-heart liberal McGovern. Obviously all that dope you smoke has fried your brains out."

Bill said, "No, Art, we aren't. You know the saying, 'Don't change Dicks in the middle of a screw. Reelect Nixon in '72!' It went downhill from there."

One morning it was time for our surveying instrument lesson. I could tell immediately by the sick grin on Art the Fart's face and his insincere pleasant manner that this was to be no ordinary government service day!

"Now, boys," he slyly said, "I'm gonna teach you how to set up this here transit over a specific spot on the ground. The transit will also be perfectly level over that given spot. Our objective is to accomplish this in less than two minutes. Now of course I expect you to take longer the first time you try, but you'd better get it done in two minutes the second time." He said the latter with a slow, deliberate sincerity.

So he took about thirty minutes showing us the sequence needed to accomplish the feat of setting up the transit, "perfectly level over a given spot." Then it was our turn to try and sure enough it took a little longer than two minutes—more like twenty minutes. Art the Fart didn't say a word, choosing instead to let the tension build. Then it was time to apply what we had learned in the field. Art the Fart found a spot for us to set up the transit, handed it to us and looked at his watch. After two minutes we still weren't anywhere close to accomplishing our task and all hell broke loose. A continuous stream of obscenities and epitaphs issued forth from Art the Fart's mouth that would have done any drill sergeant proud. "I told you dumb shits to get it done in two fucking minutes. Your two fucking minutes are up! Yer wasting the fucking government's time, my fucking time and the time you're fucking mothers were in labor. The best part of you shitheads either dribbled down your fathers fucking leg or was thrown out with the fucking after birth." This went on and on for what seemed like forever until we were done. Then he abruptly shut up.

This was to be the pattern from then on. If we didn't want a red necked asshole with bulging Rodney Dangerfield eyes, bulging neck veins the size of tree branches, and spittle in our face, we had to get whatever task he assigned to us done quickly or put up with nonstop verbal abuse. Or quit. We all felt, of course, like quitting but Larry and I were determined to outlast Art the Fart. Bill, however, was teetering on the brink. One morning he finally broke. Art asked him for a plumb bob, and Bill had committed the cardinal sin of leaving his in the truck, fifty yards away. Art the fart berated him, his family, his heritage, his politics and everything else imaginable about Bill as he went to retrieve his "motherfucking" plumb bob.

"Get a move on you, piss ass. I want to see those buns wiggle," Art the Fart spat out.

Bill stopped in his tracks. "Which way do you want to see them wiggle, left or right?"

Now Art the Fart was not known for tolerating insubordination or back talk. His veins and eyes popped out even more than usual as he screamed, and I quote, "YOU CAN JUST WIGGLE THEM ON DOWN THE FUCKING ROAD, YOU GODDAMNED MOTHERFUCKER, 'CAUSE YOUR ASS IS FIRED!"

And Bill did just that. We never saw him again. Well at least Larry and I had learned better than to give old Art the Fart any lip if wanted to keep working in Yellowstone!

Now I must admit, Art the Fart's tactics turned us into excellent instrument men in no time. Fear can accomplish wondrous things. One day, though, Art the Fart was his usual unpleasant self and something snapped in me too. Our job that day was to set a final center-line down a preliminary road that had been placed and hardened by giant earth scrapers, bulldozers and other heavy machinery in preparation for paving. One of us looked through the "perfectly level transit set over a particular spot" per Art the Fart's explicit instructions, and guided the other down the road with a five-hundred-foot tape. At every fifty feet, the tape man would stop and the transit guy would view him through the transit telescope and direct him left or right until he was dead on the center-line with the tape. A stake would then be driven in that spot to direct the pavers. But it wasn't as simple as it sounds. The preliminary road was so hard that if you tried to drive a wooden stake, it would just shatter. To remedy this problem, one had to position a large metal spike affectionately known as a bull prick on the center line a whale on it with a sledge hammer to make a foot deep hole. The wooden stake was then placed in the hole.

Art the Fart wasn't happy about the pace and let us know in his customary way. I got angry and said to myself, "You want fast, mo——fo——, I'll show you fast." *WHAM, WHAM, WHAM*. In went the stake. *WHAM, WHAM, WHAM*. In went another. Before Art the Fart could say shit, we had two miles of road done, some two hundred stakes driven in one day, four times the usual pace. Art the Fart was impressed. "Maybe the best part of you assholes wasn't thrown out with the after birth after all," he grudgingly admitted. From that day on Art the Fart never yelled at us again. He even seemed to take a liking to us, impossible as it may seem.

In fact, the whole crew accepted us when it turned out they were actually going to finish the job ahead of schedule, thanks to us hippie surveyors. We even got to goof off for a couple of weeks with Art the Fart's blessing because the surveying was basically done. And final wonder of wonders, Art the Fart TOLD the head engineer to let us have several days off with pay to hike around Yellowstone with some of those desperate, severely isolated college aged women. The head engineer knew better than to back talk good old Art the Fart.

Of course we needed to resupply with groceries so every weekend we would drive the forty-seven miles to the nearest Supermarket in West Yellowstone.

BANDIDOS OF WEST YELLOWSTONE

On occasion while working in Yellowstone, we would drive the 47 miles to West Yellowstone for groceries, visit a bar and generally have a good time on our day off. West Yellowstone is a sort of micro-nism of the worst America has to offer: 120 garish motels, numerous bars and greasy food restaurants, hookers and rampant drug use. It's located on the border of Yellowstone and during the tourist season fills to overflowing with tourists looking for a place to stay after they have driven through Yellowstone only to find each and every camp-ground filled to the gills. I don't think it is possible to fill every motel in West Yellowstone. There's ALWAYS room for one more person. The streets are full of people even at 3:00 a.m. because some of the restaurants and curio shops never closed, ensuring that some sleepless tourist could always find a place to buy a replica of Old Faithful or have a beer.

One day Bill and I (this was before he was fired or quit or what-ever) parked the Volkswagen on the street and started walking toward the local grocery store. As we passed by a fine establishment known as The Pizza Palace, we had noticed a large number of motorcycles, or hogs as they are affectionately known to bike gang members, parked outside.

Suddenly we heard a large commotion coming from within and just as suddenly this guy came flying backward through the swinging saloon type doors and landed on his back on the wooden sidewalk out cold. Then a biker came running out, went over to his "hog" retrieved a long metal chain, and began twirling it over his head as he began to reenter The Pizza Palace. Before he could do so, two other bikers came out and stopped him, shouting, "It's cool, man, it's cool." He lowered his chain to the ground. All three took turns trying to revive the poor biker out cold on the sidewalk by slapping him in the face and yelling his name loudly. He finally came to, they hoisted him to his feet and all four of them disappeared back into the Pizza Palace.

I took one look at Bill and said, "That looks like a cool place—let's go in for lunch."

"Are you nuts?" Bill exclaimed. "It's full of Bandidos!"

"Aw, they won't do anything to us if we don't do anything to them," I explained to Bill. Bill was very dubious but after much persuasion he relented and in we went.

The place was quite crowded—about a fifty-fifty mix of bikers and tourists. Surprisingly, none of the tourists had left during the ruckus, perhaps because they had already paid for their fine meals and/or were enjoying the extra attraction of witnessing a bona fide biker gang, before or after visiting boring old Yellowstone. It's hard to say which. Some people were leaving and we grabbed their table and ordered— you guessed it—pizza and beer. Next to us was seated several bikers. One of them gave us the look-over and finally shouted out, "Scooter people, eh!"

"Excuse me?" I said.

"Scooter people! You're not bikers, and I can tell you're not tourists by your blue-collar appearance, so you're in between—so you're scooter people. Care to smoke some weed?"

Now I certainly didn't want to be rude and offend this gentlemanly offer, so I said, "Well sure, if you really want to. I think there's a place out back we can go."

"Out back?" said the Bandido. "What the hell's wrong with right here?"

He then whipped out a large baggie full of weed and proceeded to roll up several joints, light them and pass them around the Pizza Palace. All the tourists respectfully declined to join in. As the place began to fill with the odor of marijuana, the bar's burly red-necked bouncer strode over and declared in a loud voice, "Hey, what the hell do you think you're doing? You can't do that kind of shit in here!"

The Bandido threw his feet up on the table, took a big, long toke of pot, and between suppressed coughs said: We'll do whatever the fuck we feel like in here, man. Cough, cough, cough." He then blew a great cloud of pot smoke right into the bouncer's face. The Pizza Palace had become deathly quiet, as all the patrons were watching this confrontation take place. The bouncer's eyes darted from table to table and I'm sure he couldn't help but notice about twenty bikers staring at him with smirks on their faces. His eyes bugged out, he clenched and unclenched his fists and jaw, and his veins bulged in his neck. He sort of reminded me of Art the Fart. Ten seconds went by and he finally blurted out "Okay, but when you're done with that one, NO MORE!" Then he disappeared back into the kitchen, probably to weep. The bandido continued to roll joint after joint until the entire baggie of weed was gone. The Pizza Palace simply reeked with smoke, the 'no smoking' signs apparently unnoticed.

Amazingly, nearly all the tourists were still present. Perhaps it was a testimony to the Pizza Palace's fine cuisine. Perhaps they were mesmerized by the entertaining bike gang. Perhaps they were enjoying the second hand high. It's hard to say, but nobody had left!

Now you may be asking, where the hell were the police? You should know that West Yellowstone's government must not have had much of a budget for there were only three police officers on the entire squad and I think one of them was named Barney Fife. Apparently they had wisely decided to go sightseeing in Yellowstone, leaving the interpretation and enforcement of the law the local bar bouncers and Yellowstone's entrance fee collectors.

The joint rolling bandido was getting thirsty from smoking all that weed, but he was short on funds. He asked us for a "pitcher contribution" and we felt obliged to pitch in a couple of dollars each but it still wasn't enough money. But the bandido had a plan. "Watch this" he said to us with a wink. He got up went over to the next table, occupied with a tourist and his wife or girlfriend and asked in a loud, overbearing, intimidating manner:

"Hey, man, we're taking up a collection for a pitcher of beer. Would you care to contribute?" The guy stammered around a bit and finally said, "Well, I guess I could spare a dollar."

The bandido repeated this procedure at several more tables, went up to the bar, purchased a pitcher of beer, returned to his table, slammed the pitcher down and proceeded to swill some more beer. Bill and I filled our pint glasses. "See how easy it is?" he said to us. "All they had to do was say why the fuck should I give you any money and I would just have moved on to the next table."

While all this was going on, more and more bandidos were arriving. Apparently they were all on a sightseeing tour to show that they weren't really different from all the rest of us. They just wanted to see Old Faithful erupt and a bunch of roadside bears too. Kind of like their annual Christmas Toys for Orphans drive. Despite their gruff unwashed appearance they just wanted to "fit in."

But they weren't able to "fit in" as more and more kept arriving. The Pizza Palace was simply too full and all the tables were taken. But the pot rolling bandido had a solution for this problem. He suddenly arose, jumped up on his table and roared out, "ANYBODY THAT AIN'T INTO SERIOUS DRINKING, GET THE FUCK OUT OF THIS HERE BAR!"

Now the tourists finally made their move for the exits. Every last one of them. After the shuffling of chairs died down, there were lots of vacant seats for their compadre gang members. Although Bill and I weren't really into "serious drinking" we for some reason didn't think he meant for us "scooter people" to leave so we stayed. The bar owner made an appearance and said the gang had had too much beer to drink and she was "86ing them"—a bar term for "no more beer for you!" The gang leader got up laid a large wad of cash on the bar and said "I'll make you a deal. I'll give you this deposit to cover our expenses and give the waitresses a real good tip. You just keep the beer flowing. If you don't, we'll tear this place apart. Maybe some of us will get arrested for it, but the rest of us will bail them out and we'll split. I doubt the State of Montana well bother with the expense of extradition to have us face disorderly conduct charges. Meanwhile, you'll lose weeks of business in the height of the tourist season. So what'll it be? Now I'd like another Bud, please.'

The bar owner picked up the money and got him another Bud.

About another fifteen minutes went by, and we watched the biker gang do what biker gangs do. Basically swill beer, play pool and fondle their mamas. The pot rolling bandido continued to roll joint after joint. My eyes caught the eyes of the gang leader at a distant table and he was staring right at me with his cold beady eyes, a Charles Manson lookalike. I looked away for several seconds, looked back and he was still staring. I looked away once more and glanced back once again. This time he took his beer mug and shattered it against the fireplace without once taking his eyes off mine. I nudged Bill and said, "I think it's time to go do that grocery shopping."

He agreed.

We returned the next weekend to get groceries again to find the Pizza Palace still intact. Apparently, the bandidos were true to their word.

YELLOWSTONE IMPRESSIONS

Yellowstone—I think of Yellowstone as North America's savanna as much as I do about being its thermal, hot spot. Herds of buffalo and elk roam through its high meadows, bears are common. Animal life is everywhere. There is a constant interplay between forests of Englemann Spruce and Lodgepole Pine and large hundred-acre-plus meadows, making it an incredibly easy place to get lost in. You can camp in one meadow, venture out on a hike and if you do not mark your trail, after passing through two or three different meadows you have no idea where you are—will you be able to guess your way back to the meadow your camp is in? The solution of course is to mark your trail with brightly colored tape. Wilderness ethics required you to remove those ribbons on your way back. What a wilderness Yellowstone is! People think of it as crowded, but if you venture off the highway a quarter mile as we sometimes did you are pretty much guaranteed not to see a trace of Man. The emptiness became a Presence, you could feel it was out there—or not out there, I guess.

A group of us actually did get lost. I was working as a surveyor in Yellowstone and on our day of we decided to take a hike to a nearby lake. After about three miles the way trail got fainter and fainter as we went and before you knew it we realized we were not on any trail. Several attempts to retrace our steps to the way trail failed. We were lost. But I had an idea. I said everyone be real quiet and listen. After several minutes you could hear a large vehicle on the distant highway from where we started our hike. We hiked toward the sound, pausing occasionally to listen again. It worked. Eventually we hit the highway about a mile from our cars.

The nice thing in winter of course is that your skis blaze your trail through the meadow mazes so unless it is snowing heavily it is pretty hard to get lost. One January a group of twelve of us went on a vacation there. An outfit there had an isolated a camp away from the main highway so once there it was dead silent. They had a series of circular canvas covered yurts connected to each other and the dining room via five-foot trenches cut into the snow. Each had propane heat. We had chef-like dinners every night—Yellowstone river was so close you could easily reach it on cross country skis and you could see Yellowstone Falls caked in ice They took us by snowcat to a different spot each day, on one trip they had to stop for an hour during a white out because they feared getting off the snowcat trail and becoming stuck in many feet of true powder snow—with us in it, not good at-ten degrees Fahrenheit. Memorable trip for sure been there done that certainly hope to do it again one day.

Old Faithful Inn. What can I say but majestic! Priceless, one of the most valued structures in all the US, right up there with Timberline Lodge on Mount Hood. Entire thing built out of, what else, Lodgepole Pine logs. No forest fire was going to burn that place down, they circled the place with fire trucks when the Big One went through there in 1988. Our surveyor camp was a mere seventeen miles, thirty minutes away and we spent many a night there after work, drinking a beer or two on the second or third store circular balcony overlooking the enormous foyer and fireplace, so big they put entire foot diameter logs into it. A constant stream of tourists flowed below us—or we sat out on the flagstone terrace overlooking the geyser when it off. A girl watching paradise for sure—not that we were doing that.

The weather there was very cold at night thanks to the 7,700-foot elevation. One morning in August it dipped to twenty-two degrees Fahrenheit. It snowed all day on the thirtieth of June. They had to keep the water flowing in the pipes starting in October to keep them from freezing and by December they simply turned them on full blast and let them run until spring. Pumped in from a lake and drained back in.

Many years later I returned to Yellowstone to do some backpacking.

RANGER STATION ARGUMENT

Back in the mid-1980s, I decided to go on a backpacking trip to Yellowstone. At about this time, a new computerized backpacking permit system had been installed. And I was to be one of its first victims. I'd like to go backpacking to Heart Lake I cheerfully said to the ranger after waiting in line for some time.

"Do you have a permit?" the ranger asked.

"No, I don't," I said.

She gave me the kind of look that seemed to say, "What kind of an idiot are you? Do you really think you can just waltz on in here and go backpacking without some sort of reservation?"

But instead she wearily said, "Let's check the computer to see if anything's available. Nope, sorry Heart Lake's filled up for the next five years."

I studied various maps and came up with a new place to go.

"Nope," she said after tapping away for a while, that's full up too. After several more tries, I got exasperated and said, "Why don't you just tell me what isn't full, and I'll simply go there." After much more tapping she found a little used trail where I could stay in an area for a whole two nights. You can stay at the Phantom Fumerole camp. The first night you'll be staying in camp A-1. The second night you must move to camp B-1 because A-1 is taken then. I was so happy just to get a spot that I didn't point out an obvious flaw in the plan—that I could simply stay in camp A-1 and the new people could have the then vacant camp B-1 the second night because I wasn't going to be in it. After all you don't dare argue with the computer. Whatever it says, you do!

So I hike in the seven miles to Phantom Fumerole camp after a very enjoyable lunch at Phantom Fumerole. I had it entirely to myself, no boardwalk and no tourists. There was no one at the camp spot either as I dutifully set up my tent at campsite A-1. The meadow I was camped in must have been at least one hundred acres in size. No one came that evening either. The next day I hiked around the area and decided to just stay put at camp A-1 instead of moving to B-1. I was hoping the people coming in weren't orthodox computer users and would have the nerve to disobey and stay in vacant B-1. But no one came and I had the entire place to myself for a second night.

As I hiked out the next day, an anger slowly built up inside of me. The rangers were saying these campsites were full and they weren't because people with reservations weren't showing up. I thought about how many people must have come to Yellowstone like me seeking to backpack but were told, "Sorry, we're full." Not only that, but vast areas were being reserved for only two to three parties, thereby greatly limiting the number of people who could go backpacking. The more I thought about it the angrier I got.

When I got back to my car, I decided to pay the Ranger Station a visit and discuss their backpacking reservation system with them.

I waited in the customary line, and when it was my turn, the ranger said to me, "Can I help you?"

"Yes, you can," I said. "You can tell me why the hell you are telling people that the backcountry camps are full, and they aren't! I just came from one such camp that was supposedly full and had the entire one hundred area entirely to myself. There must be dozens of people coming here from great distances hoping to do a little backpacking and you're denying them the chance to do so by telling them that where they want to go is full when it isn't. What kind of crap is that?" I said my voice now loud enough so that about twenty people in the Ranger Station could hear me.

"We can't help it if people make reservations and don't bother to show up," said the now-agitated veteran ranger.

The Ranger Station had now become deathly quiet.

"Well," I continued, "have you ever thought about trying to do something about it? I think the next time I come here I'll just head out without bothering to get a permit and simply hike off trail somewhere and camp since it's nowhere near full out there anyway."

"That's illegal," said the ranger, now red in the face. "You can camp in designated areas only with a reservation. When we find your car parked at a trail head, we'll just stick a citation with a hundred-dollar fine in it on your windshield."

"I bet I can around THAT," I said.

Everyone present was now raptly staring at us arguing, our faces just a few inches apart when another Ranger quickly grabbed me by the arm and pulled me into a nearby office and told me to have a seat.

"Let's have this conversation in private," he said. "There's no need to make a scene. Care for some coffee?"

I accepted his offer and cooled down somewhat.

"I agree with you," he finally said. "The computerized reservation system is new and very flawed. What are your suggestions?" I was taken aback by his interest in my opinion.

"I think you should only give out a percentage of reservations in advance and keep the rest for people who walk in. That way if there are no shows, people will still be able to go backpacking."

"That's a good idea he admitted."

"And people should be able to camp a quarter mile off trail like they can in most wildernesses."

"I can't agree with you there," he said.

"Why not?" I asked.

"Because a lot of people simply can't be trusted. They litter, cut up trees and disturb the animals, and poach."

"I wouldn't," I said. "Why should I be punished for the actions of others?"

"Maybe you wouldn't," he said, "but the way I see it, Yellowstone is for the animals. It's a refuge for them. Why should they be subject to any more human intrusion than there already is?"

Deep down I knew he was right and felt ashamed for selfishly wanting to unnecessarily intrude on the animals' home.

"But we'll see about you backpacking reservation idea. I hope there are no hard feelings," he said and offered his hand. I shook it and left.

I didn't know it at the time, but I was talking to a ranger who was very high up in the chain of command. The next time I visited Yellowstone, they had changed their reservation system and adopted the very suggestion that I had made—hold a percentage of reservations for walk-ins. The new Yellowstone reservation system became the model for many other National Parks. I like to feel that my speaking out may have played some role in the implementation of the new reservation system. There is still camping in designated areas only in Yellowstone, but now I find myself in agreement with this policy because of the explanation by the ranger that "Yellowstone is for the animals."

LIFE IN BOULDER, COLORADO

"Where's my stuff, man?"

"I TOLD you, we ain't GOT it. They ripped us off."

"Where's my money?"

"I just told you, man, we got ripped off."

I was sitting on the front porch of my student rental house one afternoon when I heard three guys arguing in the adjacent public parking lot, a black guy and two white guys. The black guy was apparently upset because he had given them money to buy drugs and now they were telling him they got robbed, tough luck, no drugs or money.

The black guy hemmed and hawed for a while pacing back and forth with his hand on his chin. Then in a move so quick it would make a cat proud he slammed his right fist directly into the nose of the guy saying he had been ripped off. He then lowered his right hand to grasp the guy's shirt collar while simultaneously reaching into his back pocket with his left hand to produce what Crocodile Dundee would call a REAL knife that was instantaneously placed against the guy's neck. Blood flowed down his chin from his broken nose and he began to squeal like a pig. "DON'T KILL ME!" The black dude transferred his knife to the right hand, holding the collar, and reached into the guy's back pocket to retrieve his wallet with his now free left hand. In it were the bills he had given him for drugs, neatly folded away from his other money.

"I'll take MY money back," said the black dude, "and the REST just for good measure." He smacked the guy to the ground, raced across the street, vaulted a fence into someone's back yard and was gone. The ripped off rip offs raced off to get the police, claiming armed robbery and I heard them as they described the crime, leaving out the part about ripping the guy off. I said nothing figuring they got what they deserved. I didn't want to take a chance of them getting revenge on me later.

Now at this time Boulder was a major distribution point for drugs. Hard as it may seem today, each night hundreds of college students and hippies would descend on a business area known as the Hill directly across the street from the CU campus. Drug vendors would ply their products openly on the street. "HASH! LSD! COCAINE!" were shouted out and there was no shortage of business. My second floor apartment abutted a large parking lot and night after night I would look out my window to see several drug deals simultaneously going down inside parked cars. You could tell by the cigarette lighters going on constantly inside to light up pipes.

This went on for months, as the Boulder Police Department was overwhelmed. Then the Feds got into it, and things changed dramatically. They drove beat-up old pickup trucks and other vehicles that would never be suspected of being police vehicles. It was easy pickings for them—like an eagle landing on a warehouse with a bunch of sea gull nests on it.

On several occasions I watched them drive into the parking lot, and park behind a car where pot sampling was going on inside. One cop would go to the driver's side, the other to the passenger side. They ordered the people out at gunpoint, handcuffed them, got the keys, opened the trunk and took out large 40-gallon trash bags full of weed. They were then placed in a normal cop car to be hauled off. A tow truck was called, the car went away, and it was on to the next bust. Before long it was just a parking lot and shopping district again.

That summer I drove to Bremerton Washington to visit a married couple that I knew from the commune, who had moved there. They happened to be hiring at the Naval Shipyard there, I applied and was hired because I was a Navy veteran. I stayed and bought a house. A few years later, I decided to take a basic mountaineering class that the local college was famous for—now the mountaineering stories begin!

5.4

5.4 is a number on the rock climbing mountaineering difficulty scale. It is the difficulty level the old timers stopped at when climbing.

They knew via a lot of bouldering practice that that was their limit. At that level they felt they could safely climb without undue risk of falling. Above that level they felt a fall was inevitable at some point. No matter how careful you were you were eventually going to throw snake eyes. The rule became the leader never falls on a climb—because he knew what his limit was. Very few peaks required above a 5.4 anyway to summit.

Then around the 1930s along came the piton. Big stink. Now the new motto was the leader can fall—because the piton will arrest the fall. That was a huge confidence builder and 5.4 was soon left in the dust. The old timers called it a security blanket, a sort of cheating allowing for injury as part of the sport, as they saw their accomplishments belittled.

In the rock climbing world that branched out, 5.4 was now considered hardly worth doing to the experienced, onward they went in difficulty, eventually up to 5.14

There became several subsets of rock climbing. Top roping in a gym, bouldering with a soft landing if you fell off, top roping on 150-foot cliffs, the top easily accessible from another direction—and lead climbing on mountains, using protection to shorten your inevitable someday fall. The last are the "it's okay for the leader to fall" guys.

So when I took up mountaineering, there were many ways to go within it. I decided to do it the old way, the leader never falls, 5.4 is okay unprotected with experience. I also did and taught a lot of the "safe" rock climbing via top roping that negates the effects of any slip—you only fall a few inches if properly belayed. What great fun and exercise, physical problem solving at its finest! As a bonus, you find that emotions—such as fear—are optional and can be controlled.

I would try any difficulty with a top rope, but I didn't want to risk injury due to a certainty of a someday fall so I did no lead climbing above 5.4

I don't begrudge the guys that do risk injury due to an arrested fall. They probably think I am limiting myself, and I am. To each his own.

On the other hand, the Rule now is that you use protection on anything 5.0 or above, maybe even on class 4, so many thought I was being reckless by climbing a 5.4 unprotected, even though that was the way the Old Timers did it. So safety becomes a matter of perception, a variable. I think they are risking injury by going above 5.4 and they think I am by climbing 5.4 unroped. The way I see it the old timers rarely fell at 5.4 and below, the new guys always will at some point above that level. Even if the protection works as planned, you are likely to fall at least ten feet and a LOT can happen in a fall that long—and if the protection doesn't work you are so screwed.

THE MOUNTAINEERS OF SEATTLE

I used their book *Freedom of the Hills* as my textbook, and it was required reading in the basic and intermediate mountaineering classes that I took at Olympic College of Bremerton. A great book!

But the Mountaineers themselves didn't seem to believe in the freedom of the hills. Probably due to liability purposes, their climbs became a bit of a regimented march done in the safest way imaginable, with a strict set of rules. I found that too much order could actually be a hindrance because it did not allow for improvising. Sure you learned to rappel safely, as long as something didn't go wrong, but suddenly for whatever reason, you could not rappel in the manner taught. There was only one safest way, to focus on others was to add more chance of failure if you used them—better to just use the safest one exclusively and just mention the alternatives and discourage their usage.

You get butterfingers and drop your rappel device. You have a backup but get it wrong while tying it in and it ends up dropping too—now it really is dangerous to continue down using another method that you don't fully understand and haven't used—too much order can cause things like that. Better to practice all variations rather than insisting on learning only one method—the safest method—well.

So rather that follow a rigid set of rules—and lose the freedom of the hills—I decided my motto was that there were no rules, only decisions based on many different factors. The saying is no matter how CAREFUL you are, sooner or later you are going to throw snake eyes, the key word being "careful," because it is assumed that you are being careful to begin with.

Following a rigid set of rules sets you up for an accident, I think, because it can lull you into a false sense of security. When snake eyes are thrown you have to improvise, and that wasn't being taught, so it seemed, by the Mountaineers. The rules will likely make you safer than not following them as well as experience, but complacency and the unexpected can negate the effect.

Another Mountaineer Club Rule was that you should not go solo, but many of my climbs were and I have to say I loved them. It builds a feeling of confidence like no other going up a major peak like Mount Moran, addicting actually, no one of less skill holding you back and no one challenging you for supremacy—your ability to keep your schedule is all up to you. Your mountaineering plan is all up to you as well. I actually think it is safer when going solo up to 5.4 because my attention could focus more on the task at hand with no people around to divide it, and I was certainly more chicken because I was acutely aware of the risks and there was no peer pressure to push me on through something I was not comfortable with. I don't mean to be negative toward the mountaineers, in the end they are a good thing, much better to be than not. I just didn't want to be told our way or no way if you joined their club and climbed with them after taking the class.

I chose to take the Olympic college of Bremerton basic and intermediate mountaineering classes to learn the craft instead, mainly because I lived in the area. They used the same book in class and taught the same thing. However, the teacher wasn't nearly as dogmatic and had a much more pragmatic attitude. For example, you learned four ways to rappel, not just the best way emphasized.

They however had an inherent problem—they were too popular and they did not have enough experienced people to assist as volunteers for the class. That meant that they relied on assistants who had taken the previous year's class and thus had very little experience, let alone leading skills, as to what kind of problem THAT can cause——

MAY 18, 1980

Mt St Helen

The following is an account of my closest call in mountaineering. It happened on May 18, 1980, a date familiar to most of you as the day Mount Saint Helens blew up. Whenever I look at a picture of the volcano with her ash cloud rising thousands and thousands of feet into the air I'm reminded of this incident.

I was a student in Olympic College of Bremerton. Washington's Basic Mountaineering course. We had already gone through the first few lessons on basic mountaineering techniques, rock climbing, and snow travel and had ascended several minor peaks in the Olympic Mountains. We normally would have been climbing Mount Saint Helens, but due to the scientist's alarm at her apparent awakening over the past few weeks, she had been placed off limits to the public. So instead we were climbing a major Olympic peak known as the Brothers. A group of about twenty of use had hiked in the seven miles to base camp, spent the night, and were now climbing in the early morning light toward the summit. The climb was led by a very experienced mountaineer named Frank and several assistants, or rope leaders of unknown experience, at least to us students. Actually, I was to learn later, the Basic Mountaineering class was chronically short handed due to the large numbers of people needed to assist an average class size of forty-five students. The other half of our class was climbing as another group on a nearby peak.

We ascended for a couple of hours toward a feature on the mountain known as the hour glass. As the name implies, it was a steep gully through which snow and rock avalanched through whenever any of it was loosened from above. We stopped for a break upon reaching it and then traversed below, rather than through it, because it was considered too dangerous to ascend directly. The plan was to out flank it on forested slopes until we were above it and then re-enter the gully above the hour glass where the slope lessened.

As we were traversing, we had an excellent view of the Cascades, which included the volcanoes Mount Rainier, Mount Adams and Mount St. Helens. Suddenly, someone shouted out "LOOK!" and pointed toward the Cascades. Mount St. Helens had just erupted. As I gazed with a dropped jaw, an enormous roiling, boiling cloud, ascended to the heavens, flattening out as it hit the stratosphere. Red lightning bolts jutted out from the black cloud.

Someone in denial said, "Oh it's a slash burn" (debris wood placed in a pile after a logging clear cut and burned).

"Man," I said, "that has to be the biggest slash burn in the history of the world!"

We watched the spectacle for perhaps five minutes as it slowly drifted north, blanketing the snows of the Cascades all the way to Mount Rainier, changing them from pure white to dingy gray. Several minutes later we heard the explosion, three evenly spaced volleys, two apparently echoes of the first.

The same Doubting Thomas, apparently still in denial, said, "Oh, it's Armed Forces Day, and they must be giving an eighteen-gun salute down in Bremerton."

"Man, it must be a nuclear howitzer to hear it clear up here," I said out loud.

After much buzzing conversation over what we had just witnessed and heard, we decided to continue on. Little did we know that the eruption of St. Helens would be an omen of things to come.

The rest of the ascent was uneventful, and we had lunch at the summit. St. Helens was still obscured in a haze and we were anxious to return to Bremerton for news reports of the assumed (for most of us) eruption. Frank, the climb leader, told everyone to start down, and he would take up the rear to ensure everyone was accounted for and together, a customary practice in mountaineering. He gave a rope leader, Jeff, temporary command of the climb. Jeff was a jock type, sort of like a quarterback put in the game by the coach. But Jeff had little experience. He had been a student in the class only one year ago. He made up for this lack of knowledge by putting forward a quarterback's "I'm in charge, now here's the plan, let's implement it" attitude.

None of the students were aware of his inexperience. Explicit instructions were given by Frank to gather at the base of a minor rock wall that required belaying the students. "Gather there and wait for me," he said. "Do not glissade [a controlled sitting slide in snow] until I get there."

I was on Jeff's rope team as well as another student, Larry. We were first down the rock wall and sat down to wait. Since there were about eight rope teams, it was taking quite some time for everyone to regather. Each rope team took about fifteen minutes to negotiate the rock wall. By the time three rope teams had regathered with our rope team, Jeff was becoming impatient. He decided to continue on without Frank.

"Okay, listen up, everybody. We're going to stay roped and glissade back down to the hourglass and regather there." He went over the glissade procedure, and I interrupted him.

"But Frank said not to glissade until he got here."

"I know," said Jeff, "but it'll be another hour before he gets here. We're just going to glissade anyway, so let's just do it."

"But we can't see if there's a safe run out. In class they said never to glissade without a safe run out. What about the hour glass? That's not a safe run out."

"Look," said Jeff who was growing impatient with my questioning his decision. "The hourglass is nearly two thousand feet below us. We'll stop glissading long before we reach it."

I wasn't sure and said so. Jeff finally lost the remainder of his patience. "Look," he said, let's get this straight. I'm the leader, and you're the student. Now just do as you're told."

"Okay," I said, "if you're sure it's safe."

Jeff sat down and started sliding in the snow. I watched as the fifty feet of slack rope played out and, when it was tight, started sliding down the moderately step hill too. Larry was expected to repeat the process. When all three of us were glissading, the next rope team was to follow suite.

I had glissaded less than one hundred feet when the avalanche started. All the snow around me began to slide. Jeff yelled up at me to self-arrest, but by the time he did, I was already over on my belly stabbing the moving snow with my ice ax. It didn't work. The snow was sliding to a depth that was deeper than my ice ax could penetrate, but suddenly it held, and I came to a stop. By this time a large mass of snow was gathering speed, taking Jeff with it. The rope became taut, and I was literally ripped off the hill and started to cartwheel.

I then went through what could only be described as the three stages of shock. First stage: disbelief.

"This is a class!" I thought. "These kind of things don't happen in classes!" Disbelief was quickly followed by panic. "Oh yes, they do, and it's happening right now!" I began to flail about, trying self-arrest several more times. I tumbled several times, tried to self-arrest, and realized the ice ax was no longer in my hands. It had been ripped from my grip. I tumbled several more times and became entangled in the rope, my arms pinned to my sides. I was like a fish on a line, being pulled under the snow and out again. When I went under the snow, it was forcefully shoved up my nose, and I nearly "drowned" before I was able to twist my head enough to stop the snow from entering it. We dropped nearly two thousand feet, reaching the hourglass quite quickly, and shot through it onto the spot where we had witnessed St. Helens erupt. But we didn't stop there. We continued over a fifty-foot cliff. The only thing that saved us was the massive amount of snow that proceeded us to cushion the impact from the fall. The third stage of shock set in: objective resignation. "You're going to hit bare rocks soon," I reasoned. "It's going to hurt like hell. Your best option is to simply black out." So I did just that. I willed myself into unconsciousness.

I awoke from the jolt of a massive tug on my body. I could see rocks less than ten feet away. I had stopped just in time. Someone behind me was screaming for help. The rope was wrapped around me so tightly that I could barely breathe, and I was hyperventilating.

"Just a minute," I said between gasps. "I'll be there soon."

After five minutes, I had caught my breath and began untangling myself from the rope. I looked behind me and saw Jeff buried up to his neck in snow. He had fallen through a thin spot near the bottom of the slope and had acted as an anchor to bring me to a stop. His arms and pack were twisted up over his head. The pack was preventing a large stream of snow from sliding down and covering his head. Larry sat nearby in total shock, hands and head on his knees. I strode over to Jeff and undid the straps of his pack, freeing his arms. Over a ten-minute period, we worked together, and Jeff was finally free.

But there was more to the story. While digging Jeff out, we could hear people several hundred feet up the slope screaming for help. A couple ran past us and admonished us for not assisting those above. "There's a guy buried up there, and they need all the help they can get. We're going out to get help." Help was at least nine miles away. We started climbing back up slope to the rescue scene, and I felt sharp pain in my right side. I figured I had suffered a sprain of some sort while tumbling. When we reached the scene, we saw two lengths of rope disappearing into the snow. The student that was buried was the middle man on the rope. Two other rope teams had started glissading after we did and were also caught in the avalanche. Several students were trying to dig him out using nothing but ice axes with little success. The snow was the consistency of sugar and kept filling the hole they were trying to dig. The group's snow shovels had been lost in the avalanche along with half a dozen ice axes. The scene was very dismaying and frantic. Suddenly, a snow shovel was tossed into our midst. A separate climbing party just happened to be passing by. Progress went quite swiftly now, and we dug down ten feet, following the rope, until we saw the hair on top of his head. It wiggled, so we knew he was still alive. After being buried for more than forty-five minutes, we finally freed the buried student. He had luckily fallen into a thin spot of snow just above a stream and then had been buried standing up in an air pocket that saved his life. He was nearly hypothermic so we bundled him up in warm clothes and heated up some tea for him.

As I walked around, I could still feel a sharp pain in what I thought was my upper leg so I began to drop my pants to have a look when I saw blood. Luckily, I was already in shock. I discovered a one-inch-wide puncture wound in my abdomen. Apparently, I had been stabbed by the ice ax in one of my tumbles. By opening it up, I could tell it went in quite deeply, perhaps even puncturing my intestines.

There was nothing else to do but hike out. It would be hours before help could arrive. After the buried student warmed sufficiently, we started out. When we reached our back country camp, other students took all my gear and carried it out for me. The puncture wound continued to ooze blood, but at least it was located in an area that moved very little when I walked. Whenever I stumbled sharp pains shot up my right side. To pass the time, I flirted with one of the women from the group of mountaineers with the snow shovel, jabbering up a storm all the way out. We eventually reached the trail head where a plethora of rescue vehicles awaited. I was still in shock and just wanted to get in my car and go home, but I didn't say so. I knew all these people had come up just to help me, and I felt obligated to let them. They checked out the puncture wound and insisted that I go to a hospital about forty miles away for further evaluation in case my intestines had been punctured. They also confirmed that Mount St. Helens had indeed erupted, killing dozens of people. A student drove me and my car to the hospital, and I checked into the emergency room. A doctor came in and inserted a cotton swab into the puncture wound for a distance of two inches. He wanted to operate immediately, but I protested. I had had enough people telling me what to do for one day.

"Doctor," I said, "it's I been nearly ten hours since I was punctured. The ice ax was probably quite clean and sterile from the cold snow. Don't you think I'd have a fever if my intestines were punctured? I don't have one."

"Maybe you're right," he said, "but maybe you're wrong. The only way to tell for sure is exploratory surgery."

"How long would it be before I healed enough to go back to work after surgery?"

"We'd have to cut through your abdominal muscles. They would take about six weeks to heal."

"I think I'll sleep on it," I said, "and decide in the morning."

The nurse put an IV in my arm and left.

The next morning I awoke quite early, and no one was around. I did not want to have to go through another round of persuasion by the doctor, so I simply pulled the IV from my arm, got dressed, snuck past the admission's desk, and drove home. Still obviously in shock, I went to work in at my job in the shipyard the next day, but they sent me home for a week with sick pay when they discovered I had a two-inch deep puncture wound in my side. The mountaineering class instructor called me up, and I could feel that he was worried about a law suit because he reminded me of the liability release I had signed before starting the class.

"I had no intention of trying to sue," I told him. "I intended to continue in the class instead. If it had been my decision to glissade, I would never climb again, but it wasn't. I did nothing wrong, and neither did your climb leader, just Jeff. I just went along for the ride, so to speak."

The very next weekend, I ascended another peak with the class. We never saw Jeff again. After graduating from the class, I was ready for Rainier on my way to summiting some 250 summits.

MY FIRST CLIMB OF RAINIER

I had taken the Olympic College of Bremerton Basic Mountaineering Class and as a graduate student. I was entitled to participate on private climb of Mount Rainier. The leader of the climb was a friend of the climbing instructor and was a cross country coach at a local high school. He basically made the five-thousand-foot climb from Paradise to Camp Muir into a race, at least from his competitive perspective. The entire group was always several hundred feet behind him. When he took a break and we caught up to him he took off again, daring us to keep up without a break for ourselves, giving us a choice of taking no break and keep up or taking one and falling even farther behind. I wasn't used to the competitive nature the climb seemed to turn into—it seemed like a waste of energy.

After a night at Camp Muir we climbed on to the summit along a well-worn track in the ice, laid down by the Rainier Mountaineering Incorporated guides. Being roped up now was essential because of the many crevasses that need to be crossed, so staying together out of necessity was the result, and we were only as fast as the slowest member. Red pennants attached to bamboo stakes shoved into the ice marked the way in case of snow or fog, placed every couple of hundred feet. As we neared the summit a group of two rope teams came up from behind that was faster than we were, led by the paid guides. One rope team decided to pass below us as were taught to do when passing someone on a steep snow slope with one main track. That was the proper protocol because that way their rope would not interfere with us in any way as they passed. However, a second rope team began to pass above us and it became clear they were in a race with the first team, trying to get by us before they did so they could then be in the lead.

More competition going on. Their rope did indeed sag down into our path, and I was forced several times to move it out of the way so I would not step on it with crampons, not a good thing for the integrity of the rope. It was also a waste of energy trying to avoid doing so. Finally I got exasperated because it was taking many minutes for them to pass us—a rope team is 150 feet long and we were moving too—and it was irritating having to manage their rope, so I yelled out, "If that rope gets in my way one more time, I'm just gonna step on it. You should be passing below us like your other team."

The guide leading the rope heard me and said, "If you do that, you will be in BIG trouble."

I started to explain why I was upset, and he cut me off, saying, "Who the hell do you think put this track here, pal? We'll do as we please on it, and if you don't like it, don't climb here!"

I of course had a different opinion, and before you know it, he was right in my face. The other guides had to pull him away before he hit me, and they later apologized for his behavior.

Later on the summit one of our group pulled out a five-pound watermelon from his pack for a group lunch treat, He had kept it hidden until that moment, carrying the weight up some nine thousand feet. So our summit picture looked surreal with the group eating watermelon pieces, so incongruent.

We went back to Camp Muir the same way, the rude guide nowhere to be seen probably way ahead of us racing down.

On the way down to Paradise from Camp Muir, when we no longer needed to be roped up, the leader of our group decided to make it a VERY fast clip down just like he had on the way up, and I got irritated again because everyone seemed to think this was some sort of racing competition. When he reached a point and decided to take a break, I just kept going when I got there. He realized immediately what I was doing, put his backpack on and began racing after me. But I simply would not let him catch up, just like he would not allow us to on the way up to Muir. I wanted him to know like what it felt like to be dominated, even though that was not supposed to be the name of the game here because it was a team effort, not an individual sport, at least on Rainier anyway. By the time we got to the Paradise parking lot we were literally running with heavy rope laden packs on, through the slushy afternoon snow—proof that he viewed this as some sort of cross country race—and I beat him by several steps into the parking lot. I don't think it endeared me any to him, I got no "congratulations."

MY CLIMBING AND GUIDING CREDO

Conrad Kain was a famous mountaineering guide in Canada responsible for sixty first ascents in the Canadian Rockies, often taking complete amateurs to the tops of difficult peaks by any standard, such as Robson and Bugaboo Spire, both major efforts. He was just that good. I read his autobiography *Where Clouds Can Go*, and I could see why he was so popular—he had a much different attitude than most guides. He was simply very gregarious and knew how to instill confidence, making it more likely for his clients to at least have a positive attitude and try—that was his secret for success as well as popularity.

He came up with the four principles for being a good guide:

1. Never show any fear.
2. Pay special attention to the weakest member of the group.
3. Be able to tell a white lie convincingly if the need arose.
4. Tell someone off, also when appropriate.

To be able to talk a beginner into doing a tough climb is not easy, even less easy is getting them up and down in one piece, but that is exactly what he did during many of his sixty first ascents in Canada.

After taking Olympic College of Bremerton's basic and intermediate mountaineering courses, I became a guide for them for many years. I used Conrad Kain as my inspiration for doing a good job of it. I used his four principles all the time, although I had to assert authority by telling someone off very infrequently.

If you show fear so will the students. I experimented once. We were descending Guye Peak near Snoqualmie Pass, one of our practice rock climbs. Everyone did just fine going up, about four pitches of easy rock to practice belays on. We decided to go down an easier way since we were pressed for time, but there was some exposure along a serrated ridge. We were no longer roped up because it wasn't necessary just class two—low three at best but no handholds just an easy downsloping ridge, cliffy on one side. I decided it was a great time to experiment so I said boy that's a long fall if you fell down that side huh. Sure wouldn't want to do that! The group was progressing just fine and before long we would be on a forested slope. Some of them began to go a lot slower with a lot more obvious apprehension, almost crawling now, simply because I had pointed out the exposure. Not a good thing especially if pressed for time, can guarantee an unplanned bivouac.

Paying special attention to the weakest member was a no brainer—you only go as fast as the weakest link and the chain of safety can be easily broken if he gets in trouble. Your job is give him confidence, not dominate him. He may just give up if you do, end of climb. Sure sometimes you get impatient but expressing your frustrations doesn't help. You must plan for him in your schedule.

Telling white lies amounted to motivation when things got exhausting. For example, you know the summit isn't at the top of the slope you are on, but you lie and say look we are almost there! Worked many times just to motivate them a little further when they got exhausted. By the time they realize you are lying, the actual obvious summit would hopefully come in view and that takes over as a motivation instead. The point being the lie got them higher up making the summit more likely.

Telling someone off—or being alpha male assertive—comes when someone in the party—say another rope leader—is continually being disagreeable, negative. It gives the impression of anarchy, of group disintegration, and when it gets dangerous you want to be as one, with only one clear leader whose judgment you trust. Saying yes or no to continuing on a climb when hazards arise, is another time you need to assert yourself if overly questioned, rather than inviting mutiny or injury.

I say if you are going to go as a team, stay as a team—it shouldn't devolve into an alpha male duel, nobody enjoys that.

I also agree with the Sherpa philosophy when it comes to climbing.

One time I was ascending a peak solo in the Olympics and I had the strangest feeling. I felt the Mountain or Place had a sort of supra-consciousness, a sum of all the living entities on it and there was something very odd going on here. That oddity was my presence, giving the supraconciousness a different perspective of itself because I, a visitor, was now a part of it, something that was not normally there. I realized that it was up to me as to what would be added to that perspective. I chose to add the concept of beauty to it, and I feel that attitude has kept me unharmed all these years—why would the consciousness be indifferent to the concept of beauty? I figured I was an asset that way, not just some stranger on a personal peak bagging mission. Call it flattery in exchange for safe passage, for the Consciousness not being indifferent when I needed to get out of jams, self-made or otherwise.

I also chose to be humble because I felt I was part of something now, and hardly the whole story. After that perception, I chose before every climb to ask for safe passage by waving my ice ax over my head three times as a sign of respect. Upon returning I simply hoisted my ice axe up and gave thanks. Along the way, my mantra was, "I love the smell of napalm in the morning smells like—the summit" (from *Apocalypse Now*).

The objective hazards were the bullets all around me as I surfed the mountain, impervious and therefore indifferent—but still respectful—to their danger because I felt protected, fated not to be a victim, and that does wonders for suppressing fear.

Let's not forget the need for respect. I say avoid pushing on in bad weather just because you are tough, because it's a form of disrespect. Sure you don't turn back at the drop of a hat, but you shouldn't be ignoring the prospects of PM thunderstorm or a predicted storm coming in and at least plan for them. Prudence should trump ego. The mountain will help teach you that if you don't get it.

Any sort of litter is bound to bring bad karma—I was VERY superstitious about that, so I left no trace except for footprints and a summit register signature. If I saw litter, it became my litter if I left it there—that was my attitude.

JANINE'S WALL

On the Summit of Mt Rainier

Of all the climbs I have been on, this one in particular stands out, one of Mount Rainier in 1990. The climb of Rainier itself was not special, I had climbed the mountain many times before.

I was climbing with the students of Olympic College of Bremerton, Washington's basic mountaineering class after their successful completion of the course. I did this for many years as a sort of unpaid bonus to the students. Along on this climb were an engaged couple named Dan and Janine. Janine was one of the star students of the class that year—a natural rock climber along with a very endearing personality whom everyone enjoyed being around.

We took one of the normal, well used routes up Rainier, the Camp Shurman/Emmons glacier route. After a two day approach, we arose at midnight and began the final 4,500-foot summit ascent. Things went pretty normally until we neared the summit—just slogging along through the snow on a forever slope with a taut rope between us in case of crevasses and with boots in crampons to give us traction on moderately steep frozen snow.

Now Janine was a wisp of a woman and probably weighed all of 105 pounds. She was in great shape after completing the rugged thirteen-week mountaineering course, but as we neared the fourteen-thousand-foot level, I witnessed something I had never seen before or since: someone hitting the wall. We were simply trudging through the final four hundred feet below the summit when Janine suddenly just fell flat on her face. She then rolled over, sat up and placed her face in her hands, elbows resting on her knees.

"Janine, are you okay?" I asked.

No response. Her fiancé shook her shoulder and repeated, "Are you okay?"

Once again no response. We tried several times to rouse her, give her water and food, and motivate her, but she simply just sat there completely unresponsive. Dan finally asked her if she would mind waiting for us while we continued to the summit crater, a mere quarter mile away, maybe four hundred feet above us and within sight We had passed all the danger zones with nothing but a final gently sloping snow field without crevasses now between us and the summit. She finally responded with a slight nod so on we went, after making sure she was warm and comfortable in a sleeping bag. I promised we would be back within an hour.

After a few dozen steps we passed a couple on the way down, and I said, "Do you see that woman sitting down there? Could you check on her and give her a few words of encouragement? She just hit the wall and dropped in her tracks." They said sure and several minutes later I looked back to see them talking to her.

I thought back to my first major climb many years ago on Long's Peak in Colorado. I had gotten to within a couple of hundred feet of the fourteen thousand footer's summit when I was overcome with nausea and had to stop. A couple passing by had noticed my distress and stopped to offer encouragement. "Here take some of these Rolaids," they said. "They seem to help." Then they continued on, but as they did one of them paused to say, "*You know, if you don't continue on to the summit from here you will regret it for the rest of your life.*" I thought about that for a while, realized that they were right, and continued on.

Now I was regretting not saying the same thing to Janine. But the people on Long's Peak were strangers, people I would never see again. I didn't want to take the chance of Janine being resentful toward me for pointing out the consequences of her failure, during her moment of doubt and pain.

Twenty minutes later we were on the summit, with Janine's inert figure in plain sight down below. We had a quick lunch, all the time looking down at Janine and wondering what was going through her mind. All thoughts were on her, and we all felt badly that she had put out so much effort only to be stopped so short of her much cherished goal. Finally we decided to start the descent. As we gathered up our gear, Janine quite suddenly jerked to her feet and we watched for twenty minutes as she staggered the final quarter mile to the summit. Now I must admit that tears welled up in my eyes as I hugged her on the crater rim and said, "Well done, Janine, well done." I've never been so proud of anyone in my life.

We began our descent, and Janine soon returned to her normal cheery condition. "You know, Janine," I said to her. "I've been thinking about what you just did. You are the one person today who gained the most from this climb. You hit the proverbial wall, got up and went over it anyway. Now you know for certain what your reaction will be when adversity comes into your life. None of the rest of them has gained that knowledge today. They are still left to wonder what their reactions will be when they hit their walls."

I've been up Rainier via the Emmons route several times since then, and now make a point to proudly stop near the summit to point out a feature unknown and unseen by the rest of the group. It's not shown on any map, but it is indelibly printed in my mind: Janine's Wall.

MOUNT RAINIER IMPRESSIONS

Put bluntly, it doesn't want you there; it is not a place to live in, just to visit briefly. There is nothing but rock and ice in the no-life zone. Beautiful no doubt, but you would go blind without sunglasses on a sunny day, your skin would get blisters without sunscreen protection and you would freeze to death on just about any night on the summit with or without a coat. No one dares to pitch a tent in the crater, you want to get DOWN long before sunset. A simple on shore push off the Pacific Ocean in June—that barely creates a breeze in Seattle—will create one hundred miles per hour steady wind on the summit that will tatter your tent in no time.

On the other hand, if you do it when conditions are right, usually from May to June, you can do a glissade—a fancy mountaineering term for sliding on your butt in the snow using an ice ax for a brake —down from Camp Shurman at eleven thousand feet all the way down to snowline at Glacier Basin at 5,500 feet a drop of over a vertical mile in just a few minutes. At first you see these tiny dots way downslope that are parties climbing up the Inter Glacier, and then they become tiny dots way above you as you continue to descend. Been there, done that, as fun as it gets, beats any amusement ride.

And you must struggle. I tell people climbing Rainier is the hardest thing you will likely ever do. The only way to train really is to do other lesser peaks first. Of course, if you are guided you don't have to carry as much, but even then the attrition rate is remarkable 50 percent of all those who attempt it. Altitude sickness is a factor as well as weather, but the prime reason is plain old exhaustion that saps the will. Attitude is also a big factor. I have seen people turn back at the drop of a hat, saying who really cares, to people continuing on in a sixty miles per hour steady wind, to ones literally crawling to make the last few hundred feet, vomiting as they went, Determination is all they know.

Bronka is a humbler. She is a woman who at age seventy-eight climbed Mount Rainier in one day, a feat that astounded her guides, they of course had never seen anything like it. At least 98 percent of climbers take two or three days to complete the feat, very few people even in their twenties can do it in one day. Sorry macho guys, not fake news, reality! She's an inspiration too, who wouldn't want to be that physically fit at seventy-eight! Her secret was that she lived very close to the Park, and made the five-thousand-foot climb from Paradise to Camp Muir once a week for exercise.

I had a woman aged sixty-five on my rope team for one climb and she had no trouble at all, partly because I was using the "We are almost there!" white lie. Yet I have also gone with marathoners and the rope was tight as they continually needed to pause near the summit, while I wished to push on. You just never know.

I usually allowed three days to climb, depending on the route. I found that way less people get so exhausted they have no desire whatsoever to get out of a warm sleeping bag at 1:00 a.m. after zip for sleep, gear up and start climbing up a steep hill in total darkness sans your head lamp. Did I mention it is cold? The thermometer says fifteen degrees, and even though it's late June, it takes willpower just to get up and, trust me, you are not getting up if utterly exhausted the day before.

One can get grumpy real easily until things get moving. One day down from the summit was the general rule, the anticipation of a good meal in a restaurant followed by a good night's sleep is the incentive that makes it happen.

BAD THINGS HAPPEN IN THREES

Mt Cepheren

It was the summer of 1983, early in my mountaineering career.

"What's the name of this here mountain again?" said the ornery-looking eighty-year-old woman as she hiked on the nail to Chephren Lake.

"Mount Chephren," was the reply.

The "skeeters" (as she called them) were bitin' pretty bad.

"You mean sufferin'?" she joked. "'Cause we're sufferin' pretty good."

"Got that one right," I thought as I passed three eighty-somethings on the trail during my return from a fourteen-hour climb of the mountain named for a pyramid in Egypt. I had nearly been killed three different times during the descent. This was my second time here. The previous summer I had attempted a climb, but had run out of time. I made the mistake of using a sixty-year-old climbing description. The snowfields along one side of Chephren Lake were no longer there, and I had wasted much time bushwhacking through the thick brush that had replaced them. This time, I knew what to expect and was able to save hours by hiking on the other side of the lake.

I won't bore you with the details of the ascent. It was basically a lot of steep slopes, some class three to four pitches and long-seven thousand feet of elevation gain long. I was surprised when I signed into the summit register to find that Chephren, a major peak right along the Ice Field Parkway in Alberta, had been climbed only a handful of times over the last twenty-five years. Perhaps others had made the same mistake with the brush. The view, needless to say, was quite panoramic with a seven-thousand-foot exposure on its north-east side. The time came as usual for the inevitable descent. I made the decision to take a different route down in an effort to save some time. Traveling down on moderate snow slopes is always faster than descending talus and steep grassy slopes. I had spotted one such snow slope during the ascent that I thought would do the trick.

As I approached it, I noticed what appeared to be a minor *schrund* (a sort of crevasse) in the middle of it. When I reached the base of the last class three rock pitch, I gingerly stepped out onto a snowfield, a transition I have made hundreds of times in my mountaineering efforts. About a foot of open space separated the snow from the rock, something not unusual at all. This time, my weight caused the snow to collapse into an undercut cavern. The edge of the collapse was perhaps three feet in front of me and extended to my sides for many feet. My immediate (and I do mean IMMEDIATE) reaction was to push up and fall forward with all my might. I was just barely able to catch the lip of uncollapsed snow with my chest and slammed my ice ax into snow, my lower torso and legs dangling in air. I pulled up on the ice ax with such adrenaline that I slid down slope for yards.

Meanwhile, the collapsed snow fell thirty-some feet into the cavern with a tremendous KATHUNK! I probably would have been crushed (or at least busted up in the fall) by the large mass of snow if I had gone in with it. I literally shook it off and went on (like, what other choice does one have?). Next, I approached the side of the main snowfield I planned to travel down. I noticed that it was much larger than it had appeared from far above. From my vantage point, I could no longer see the schrund in its middle.

The snow was icy but blanketed with pebble-size rock debris. This gave it good traction, much like sand on a snow-covered road. Or so I thought. So I neglected to put my crampons on since it was time consuming and the snow seemed soft enough not to need them.

After traversing around two hundred feet toward the middle, I was shocked to see that the now visible schrund was also proportionally larger than it had appeared. Much larger! I changed direction and started down toward the schrund. As I got within twenty feet I saw to my dismay that it was about five feet across. I decided to turn around, but at that very moment, my feet slid out from under me and I fell onto the ice, once again slamming my ice ax into the slope. This time, it went in about a quarter of an inch because the snowfield went from soft near the edge to near glare ice toward the middle. It was still covered with small pebbles and scree debris and thus making the transition indiscernible until you got there.

My life flashed before my eyes, it really does happen—probably because I was certain that I was about to slide into the gaping crevasse only yards away. So there I was, pinned to an icy slope, my ice ax barely scratching the surface. There was only one thing I could do.

At the count of three, I abruptly pushed myself to a vertical position, hoping against hope that friction from my boots would keep me from sliding. They held. I immediately and frantically hacked at the ice with my ice ax to chop an indentation and stepped into it. It seemed too awkward, too long and too dangerous to retrace my steps, so I chopped a series of steps down to the edge of the schrund and peered down. It was so deep I could not see the bottom.

The opposite lip was five feet across from me and about twelve feet down. I felt I had no option but to jump it. I carefully chopped out a platform from which to launch and wondered about the hardness of the snow I was about to jump down on. Would I break my ankles when I hit it? (Assuming, of course, that I made it across.) It took a while to gain the courage but I finally counted to three again and jumped. I fell in slow motion and barely cleared the gap, sinking up to my knees in. Luckily, soft snow. I was now physically and emotionally drained but filled with a sense of relief. "Thank God that's over," I thought to myself. The rest of the down-climb will surely be uneventful. I started off once again down the snowfield heading for its edge recounting my close calls when I suddenly fell into a hidden crevasse. My arms shot out and arrested my fall, my feet once again dangling in midair. Right about then I lost it. I dug myself out in a panic, and stood up thinking, "It's trying to KILL me, the mountain's trying to KILL me!" Since there was no way to know where more hidden crevasses might lie, I just picked a direction at random and ran for it. Eventually I reached the side of the snowfield and returned to Chephren Lake the way I had come up, something I should have done to begin with.

Many years later, I still have bad dreams about "Sufferin'." I incorporated my near misfortunes into a new consciousness of how alert—consciously and unconsciously—to danger I had to be if I hoped to not have a repetition of my misjudgments, and to never take ANYTHING for granted in mountaineering, especially appearances.

As for my guardian angel, I think she must have complained big-time about all the overtime on Chephren.

MOUNT SHASTA

I've had four shots at Shasta and have only been successful once. The first time I went with two friends and we decided to go up the usual route from the hut. The snow turned out to be way too soft and I ended up soloing Shastina instead, post holing most of the way.

I suffered the worst sunburn of my life on that outing. It was foggy and I did not know at the time that the fog actually amplified the effect of the sun's ultraviolet rays. I did not put enough sunscreen on until it was too late and the result was a blistered and blackened face. The skin cracked and had pink lines in it. For about a week people hurriedly looked away when they saw my face because it looked like I had been permanently disfigured in a fire. About a week later my entire face peeled and my skin felt was like it was fresh and new. Almost like a chemical peel, I guess.

The second time I went with two people and only made it as far as the Mazama hut due to a snowstorm.

The third time a I was as determined to bag it as I have ever been to bag a peak, it's a long drive from Seattle—and I was in great shape to do it. I turned it into a one-day climb and had my biggest one day elevation gain ever, some 7,500 feet. I met a man collecting red snow samples for a study, who wanted to climb with me, but there was no way he could keep up, I was on a nonstop mission, taking maybe ten minutes for lunch an no sit down breaks at all.

I came upon and passed many groups of climbers who had started from Lake Helen and were still having a very rough go of it.

One rope team—you really didn't need to be roped up—no crevasses anywhere—idea of how to climb a moderately sloped snowfield was to make great looping switchbacks that gained maybe twenty feet per switchback. That approach was taking way too long.

It was obvious they would never make the summit. Another's was to push as hard and as fast as possible until exhausted, then resting for a few minutes—and then repeating the process before becoming exhausted again. Wrong again! I just went straight up the hill non-stop at a VERY steady pace, nobody holding me up, nobody pushing me.

And wouldn't you know it a thunderstorm popped up just as I neared the final summit haystack. Shasta seems to attract them. But I was defiant and determined after putting out so much effort and ascended it anyway. It soon passed. The numerous glissades on the way down were exquisite, and saved a lot of time and effort.

The fourth time I went with three others and we camped in the snow at lake Helen. Shortly after starting out the next morning, we looked back to watch the wind blow away one of the tents and Ted, who owned it, left to find it and said the climb was over for him. Irma got exhausted quickly and had to turn back to camp. I continued on with Nancy and we after passing through a notch in a ridge began the final ascent on Broadway—a ledge about fifty yards wide that is an easy scramble, but on either side are sheer cliffs. My mountaineering intuition kicked in and I got my trusty compass out to take some bearings so that we would stay on Broadway if a fog rolled in and not wander over a cliff—and sure enough it did.

As we neared the summit, you couldn't see a thing. I knew we were pressed for time if we wanted to make it all the way to the road by dark to meet the others, so I told a little white lie and said "look I think that's the summit" as a small crag appeared out of the fog, but I knew for certain that it wasn't. So I pretended to summit with her and down we started. Suddenly I heard two voices yelling "HELP!" out of the thick fog. I yelled back and soon we found each other. Two guys had become completely disoriented in the fog and had no idea how to get back to the keyhole notch that easily led down to gentle slopes. Because I had taken those compass readings we hit that keyhole dead on, and down we glissaded, making the road fifteen minutes before total dark. The catfish dinner in Shasta City was delicious, and to this day Nancy thinks that she bagged Shasta—well close enough anyway with only the haystack left to go.

Rainier and Baker have a similar situation. On Baker the true summit is a half mile away after you top out in the crater after ascending the Roman Wall. Another haystack on the summit situation. On Rainier the summit register is a quarter-mile walk across the crater when you come up on the Camp Muir side. Some people insist on the true summit while others call it good at the crater rim—mainly it depends on just how exhausted you are.

ROUTE-FINDING SKILLS

I honed my route-finding skills in the Olympic Mountains first by taking a class and then by bagging one hundred peaks there over the years. Although certainly not noted for rock climbing due to the nature of the rock, mostly basalt and shale, I can't think of a better place to learn route finding except perhaps the Tetons. I lived closer to the Olympics so I went there instead.

Route finding was important because of the factor of time in your mountaineering equation. Nothing ensures success like an early start as the saying goes, but route finding boo boos can negate that early start. Make more than one, and bye-bye summit without spending an unplanned night on the mountain, a big embarrassment in the mountaineering world they call being "benighted."

The mountains of the Olympics have no straight route up. Many cliffs crop up with slopes to the left or right that lead around them—and so on as you went up for thousands of feet on some forested slope to reach the alpine level. No trail up 95 percent of them after a backpack in on a river bottom trail to the starting point, so no help there. I learned what to look for when scoping the route up. After all, you had a choice, left or right. Guess wrong, and it's forty-five minutes shot, at least, as you must back down and then go the correct way. I played a game of no guide book, just topographic maps so that my ascent was more like a first ascent—just winging it helped a lot too in learning what to look for. I did note the time it took to get up in the guide book. I tried to factor one forty-five-minute mistake into the climb time, and still have a decent daylight cushion to get back to camp or the car.

The summit blocks on many were an absolute maze of steep gullies that you had to zigzag between on the way up to avoid cliffs, making route marking using cairns, or small rock piles, an absolute necessity. Where and when do you place them so you can find the way back in a fog if necessary?

The Olympics also had a lot of steep snow slope climbing that helped hone your ice ax and glissading skills.

The major peaks of the Tetons—the Grand, Mount Owens and Mount Moran—have no simple straight routes up them either and what I had learned in the Olympic Mountains helped a whole bunch there,

The margin for error is much less there because all involve fourteen hour days of near constant motion—6:00 a.m. to 8:00 p.m. with one-hour daylight safety cushion. A single forty-five-minute route finding screw up knocks it down to fifteen minutes safety cushion.

I only came close to being "benighted" on a climb of the Grand Teton. The group got summit fever and we didn't make it to the top until 3:00 p.m., thank God there were no thunderstorms that day. Once we were all down to a safe trail it was every man for himself as we raced to get back to camp before dark. It was tricky because we had to cross a bridgeless creek to get back to camp. An hour and a half later my flashlight's extra batteries went dead as we neared the crossing, but hers was still barely functioning, just enough to see our way across. Others in the party also caught by darkness were not so lucky. Their flashlights also went dead and they couldn't find where the creek crossing was, let alone cross it in the dark. They had to spend the night sleeping on the trail, until first light— about six hours, the price of summit fever.

WHAT CLIMBING RATINGS MEAN TO ME

Class 1—A walkup.

Class 2—You might need to use a hand for balance—steep hill or ridge.

Class 3—You could injure yourself if you slip, bruises, cuts, scrapes.

Class 4—Free fall can be expected if you slip but minor, could break a limb. Jagged face or very steep nearly vertical ridge, plenty of large foot and hand holds.

5.0 Free fall will likely kill you, still lots of big foot and hand holds.

5.4 The foot holds and hand holds are much smaller than 5.0, perhaps hand-size, and much less frequent.

Above 5.4 the footholds become more vertical and you are relying on friction as much as up and gravity to keep your foot in place, as you smear your boot on them and trust that they will hold.

And up from there to imperfections and tiny nubs on near vertical cliffs and overhangs. There are no real handholds eventually as you move up the scale of difficulty, the arm basically is used to balance you on the cliff by pushing against it, the friction from your shoes the only thing keeping you from falling.

BREWER'S BUTTRESS SERENDIPITY

Castle Peak

I had been on a roll. I was in my early forties and undoubtedly in the best shape of my life. I had gone to Canada on a two-week climbing vacation, and after an attempt at Mount Robson (close but no cigar) and successes at Roche Miette, Edith Cavell and Chephren I was looking for more to do. I went to the Canadian Alpine Club in Banff to see what they might have to offer. They had a number of huts for rent and suggested the one on Castle Peak. So I paid the four dollars a night fee, and off I went. For some reason they neglected to tell me there were several short class four pitches to navigate just to get to the hut. But they were on excellent clean rock and I made short work of them even with a full pack.

I spent the night in the hut alone and awoke without a clue as to what I was going to do that day. All the climbs out of the hut were fifth class and I felt a little intimidated doing them solo.

For a while the weather looked like it was going to decide the day for me. An early morning thunderstorm went through and it started raining. I looked out the door and spotted four people on their way up to the hut. After several minutes they were at the doorstep and asked my permission to come in out of the rain since they did not have a reservation. I said sure, of course, and so they huddled inside the small hut that consisted of four bunks and a table. I offered them some of the hot water I had, and we all proceeded to exchange the usual small talk pleasantries. It turned out they were two Canadian alpine guides in training and they were taking their girlfriends—also climbers—up something called Brewer's Buttress for a practice guide and climb. I told them what I had been up to and they seemed duly impressed. I told them I was not at all familiar with the routes up Castle Peak and was open for suggestions, but they discouraged me from trying the fifth class climbs solo.

So I impulsively said, "Well, can I go with you guys?

There was dead silence for about fifteen seconds.

Then one of them said, "Do you have a brain bucket? (Meaning hard hat.)

"Yes, I do!"

"How about rock climbing shoes."

"Um, no all I have is my Red Wing hiking boots." They hemmed and hawed saying things like I don't know if you should be going up the route in what they called "snow boots"

I finally interrupted their discussion to say, "Aw come I can do it, just give me a shot!" They acquiesced and after the thunderstorm moved off and the skies cleared off we went about a half mile on a broad long bench to reach the start of the climb. They asked me if I wanted to lead the first pitch. I deferred of course, saying, "No, go ahead, you've got more experience on high fifth class rock than I do in my snow boots."

So to make a long story short, we ended up doing some twelve pitches of 5.4 to 5.8 climbing, up some 1,200 feet. I was able to do all the pitches except for one move where I was forced to use the rope as a handhold despite their protestations because the "snow boots" were just not flexible enough to use the available foothold in one spot.

At one point one of them declared, "I can't believe this guy is doing it in snow boots!"

So they deliberately went off route and asked me to try different moves to see what the limits of snow boots were. Turns out it is about a 5.8 max.

After the descent they invited me to stay at their place in Lake Louise for the night. Seems they were quite popular at the local bars we visited that evening. The next day they decided to go to their private guide practice area at the back end of Lake Louise and we spent the day doing a number of 5.8, 150 feet pitches, much to the tourist's delight.

It turned out that the guide book said that Brewer's Buttress was considered possibly the finest rock climb in all of Canada. Talk about serendipity!

MAROON BELLE

Maroon Belle

I wasn't sure of what I had just seen. Did I just witness a car hit the median barrier on the opposite side of the Interstate? I was on my way to Denver one early summer morning to visit relatives, and had planned to stop and climb South Maroon Belle, a Colorado Fourteener near Aspen.

As I watched in horror the car hit the barrier again. This time it flipped over and rode down the concrete barrier on its top, a stream of sparks streaking behind it. Then it fell off the barrier and proceeded to roll. And roll and roll and roll, several times end over end. The trunk hood, engine hood and one door ripped loose followed by one tire and then another. Glass spewed in all directions. The crushed hulk finally came to rest just yards from where I had braked to a stop. The first thing that came to my mind was that no one could possibly have survived such a devastating crash. For about a minute it was just me and him. I was shaken so badly by what I had witnessed that I was frozen in shock. Then a trucker came down the hill, coming to a stop amidst the scattered debris. Simultaneously, we left our vehicles and walked toward the shattered remains of what had been a car. I did not look forward to seeing what I was certain we would find—a crushed and bloody body.

As we neared the wreck, I noticed movement from within. Could someone still be alive? To my amazement, the only door left began to pry open, and soon the driver emerged. He did not have a scratch.

"How on earth did you survive that?" I repeated several times. "You must have rolled ten times. Are you certain you're all right?"

He was obviously dazed, but otherwise, he seemed fine. He was a college student on his way to California, he had been driving all night and had simply fallen asleep at the wheel. Luckily for him, he had been wearing a seat belt.

"Look at my poor car!" he moaned. I tried to comfort him.

"Hey, the car doesn't matter. It can be replaced. It's a miracle you weren't killed or badly injured."

Then he panicked and wondered where his treasured baseball card collection was. I helped him find and pick up the cards lying among broken, scattered luggage that marked his path of destruction. The trucker radioed the police. When they arrived they took over and I continued on my way. What's this got to do with a mountaineering story you ask? Well, just read on.

I had attempted Maroon Belle before but was turned back by treacherous snow conditions—I kept falling through thin snow covering talus slopes and had nearly twisted my ankle several times. I don't really like risking injury in inaccessible places, so I reluctantly had turned back.

The access road from Aspen was closed by a road block about halfway up to the trail head to Maroon Lake.

"The road is closed to vehicles in summer," said the attendant, "to allow mountain bikers to use it without fear of accidents. You'll have to go back to Aspen and take the hourly shuttle instead."

This, of course, meant that I would have to forget about the climb because the delay would mean too late a start. One has to make the summit early in the Rockies due to the thunderstorms that invariably build up in the afternoons. This is a state highway. I thought and should be open to all vehicles, not just bikes. Before I could complain, she said, "Unless you have a campground reservation."

"I do," I immediately lied.

"Okay, then you can proceed. Check in with the campground host when you get there." I drove on up and, of course, went to the trailhead instead of the campground and quickly started out before I could be caught in my lie.

This time it was later in the season and nearly all of the snow was gone from Maroon Belles slopes. It was a classic class three to four ascent, first across the face of the limestone peak, and then up a couloir, invisible from below, that ran for several thousand feet, then once again right up the face. An enjoyable climb for sure. I had lunch at the summit and noticed the usual afternoon thunderstorms build as they often do in the Rockies. Best not to linger, I thought and headed on down. If you've ever been in a thunderstorm above timberline, there's no need to explain my fear of being caught in one. I made it down to the top of the couloir before the storm hit.

Lightning bolts began to shoot about and the rain came down in horizontal sheets. I tried putting on my rain gear but it was so windy I couldn't. I ducked into the lee side of a large rock, finally got the gear on and decided to wait out the storm. Several more lightning bolts hit nearby followed, of course, by loud claps of thunder. Suddenly, I heard thunder that didn't quite sound right, maybe because this time the ground shook. I noticed movement to my side and then watched in utter disbelief as two enormous 2 story house sized rocks crashed down mere yards from where I stood. They both stood on edge for what seemed like seconds before toppling into the large couloir that I was about to enter. They shattered into thousands of pieces, some large and some small and scattered down the couloir like some gigantic shotgun blast initiated by Thor. Now it was my turn to shake. Uncontrollably. The force of the two rocks had been so massive! If I had been anywhere in that 2,500-foot couloir, I surely would have been killed by rock fall. Every square foot must have been struck. It took some time for me to continue on down, for I greatly feared that more rocks would follow and I would be vulnerable for some time to come.

I thought back to the accident that I had witnessed. If I had not stopped for it, I would have been in that couloir when those giant rocks were loosened by the thunder. Many of you probably don't believe in Guardian Angels, but I do now. Especially when you consider that this person had survived, without a scratch, a crash that should have been fatal. Because of this and many other close calls, I have become a fatalist when it comes to climbing, and feel is not my fate to die on the mountain or in any other accident for that matter. This belief has calmed my fears many times while climbing. I'm just certain that it will turn out fine when I get into unexpected trouble, but I wonder sometimes if my guardian angel grumbles to God about working overtime.

WIND RIVER RANGE

The thing that struck me most about this unique place was the *looong* forty-five-mile dirt road access in, the incredible number of lakes, the barren rocky landscape that covered many square miles above timberline, and Parry's primrose.

The lakes—there were basically three kinds—ones below timberline ringed with firs and spruces, timberline ones in moss and wildflower covered terraces and still higher up the austere alpine ones with snow slopes still leading into them in August—not much but lichen living in the rocks around them.

Parry's primrose was the dominating wildflower, its clumps of pink blossoms everywhere at timberline and above, along with less common blue Rocky Mountain Columbine and the usual paintbrush and lupine

The lakes had been stocked by the Wyoming game and fish department, and they published a map showing when they were stocked and with what type of fish. One could hire a horse pack train located in the area to take you off trail to one of these many hundreds of lakes if you weren't into hiking there and the odds of seeing another party would be nil. Or just go by compass to one picked out of a hat to ensure solitude, no brush at all and terraces with easy access between them. Sounds very interesting if your thing is horseback riding, cross country travel, lake camping and fishing.

I wasn't there to fish however. I was there to bag peaks. My three trips were to Cirque of the Towers, Titcomb basin, and to Gannet peak the long way.

Titcomb basin was well used, but not overused with tents scattered near the shores of the numerous lakes in established camp spots. Fremont and Kit Carson had camped here while passing through in the mid-1800s and he climbed the nearby peak now named for him—just a scramble but what a view of most of the southern Wind Rivers, so many lakes tucked in here there and everywhere on terraces that emptied into lower terraces and their lakes, fifty feet down—and so on. That was all we did on my introductory trip there.

Another trip was a two-day backpack in to Frank Lloyd Meadows to climb Gannet Peak, high point in Wyoming—the length of the hike in kept the weekender riff raff out. It was memorable because of the moose there, wading in near-timberline lakes and the thunderstorm that drenched me just after I got down from the summit. My camp survived the downpour nicely. Thank God for rain flies that work! The glaciers there were a surprise, not visible until you gain some altitude during the main ascent. I was expecting glorified snow fields but they were the real deal just like in the Cascades, quite wide with crevasses near the tops. It felt almost like I was in the Olympic Mountains as I traveled down the lower part of one on the descent. Never saw another soul for five days.

But my most memorable trip was to the Cirque of the Towers, where I climbed Pingora Peak, considered the premier rock climb of the region—and that led to another close call.

THE RAPPEL FROM HELL

Pingora Peak

This story takes place in the Wind River Range of Wyoming.

Once again I was on my way from Seattle to Denver to visit relatives. I had heard stories about the Wind River and decided to make a backpacking/climbing trip to see for myself. I decided to visit the Tower of the Cirques and climb Pingora Peak.

I made camp near Lonesome Lake near noon of the second day in, and decided after lunch to ascend a peak called Camel's Hump, an easy walk up. One must pass under Pingora while going to Camel's Hump and it looked unclimbable, a sheer 1,200-foot solid quartzite cliff. I wasn't intimidated easily by foreshortened perception, so the next day I started out with the attitude of just taking it one step at a time. I ascended a buttress below Pingora by going up an easy slope to its left. I soon came to the cliff that looked so improbable from below and discovered Pingora's secret—a five-foot-wide ledge that ascended up the cliff at a steep angle. It was rated at class 5.4, at the limit of free climbing without artificial protection. You simply could not tell it was there when looking from below.

The rock was as good as it gets, the handholds might as well have been metal, they were so sound. I truly enjoyed the ascent, as the handholds turned into footholds, passing several rappel platforms along the way. I even found a "friend" on one of the platforms, a piece of expensive rock climbing gear usually wedged into a crack for belaying purposes, not as a rappel anchor. It took half an hour to pry a fifty-dollar piece of gear out because apparently it had been jammed in tight thanks to the weight of the people rappelling off of it—and just left there.

After the usual summit lunch, it was time to start down. Since I was soloing, I was carrying only rappel gear, not rock climbing ascent gear. I did several rappels and soon was back to the buttress. It was still fairly early so I decided to do a few more unnecessary rappels by going straight down the buttress instead of going around and down the easy way I had come up, because I was enjoying them so much.

The first two went very smoothly. I wrapped slings around large boulders and then the rope through the sling via a rappel ring (to make the rope run smoothly when you retrieved it). I tied a large knot in one end of the rope too big to fit through the anchor ring and attached a much smaller parachute cord to the knot After rappelling down with the large knot pressing against the rappel ring as your anchor, you pulled the parachute cord—attached to the knot and the rope then slithered down through the rappel ring as you retrieved it. Basically instead of doubling your rope through the rappel ring and getting only seventy-five feet out of a 150 foot rope for the rappel, the parachute cord enables you to use a seventy-five-foot rope instead and still get a seventy-five-foot rappel—thus saving a lot of backpacking weight.

But on the third rappel I did something pretty dumb. I didn't make certain that the rope went all the way down to flat ground. I could not see the where the bottom of the rope was. Well, I judged the height of the cliff badly. What I thought was less than seventy-five feet was in reality was around 125, it turned out. I'll let you do the math.

Fifty feet into the rappel, I realized that I wasn't going to make it to the bottom, because I could now see the rope dangling below me, and stopped. I briefly considered climbing back up the rope but feared getting exhausted and then free falling back down the sheer cliff. I was starting to get frantic. I noticed a narrow, tilted shelf/ledge about thirty feet to my right and I decided it was my only hope. I started to sway like a pendulum and was able to eventually land on the shelf, about the size of a coffee table, maybe three feet wide and six feet long, it jutted out from the cliff as one piece of rock. Then I pulled my rope down to me from the rappel ring above, using the parachute cord.

So there I was, stuck on a tilted ledge on a cliff. I noticed a hair line crack where the ledge met the cliff, but it was far too narrow to slip anything into it for an anchor to rappel from. At one end of the horizontal crack was another one running vertically perpendicular to it. I put my fingernails into in and pried. A piece of rock the size of walnut came loose, creating a small cavity and exposing another small cavity behind the horizontal crack. I put a sling into this cavity, but could not find anything to hook it on. The friend I had found was too large to fit into the cavity. But luckily I had saved that walnut-size rock, the only one on the entire debris-free ledge.

I put the sling back into the cavity and crammed the small rock into the cavity in front of the sling. Then I pulled on the sling which jammed the rock into the sides of the cavity thus anchoring the sling. But I had to keep tension on the sling or the rock would fall loose, turn sideways and pop out. I'm certain the Mountaineers would have frowned mightily on this jury-rigged, extremely questionable anchor, but it was all I had. It took over a half an hour of tugging on that little rock to ease my fear that it wouldn't fail me during the rappel. I finally got the nerve up, set up my harness for the rappel, prostrated myself on the ledge, and then slipped off it. All the while I kept pressure on the rope. Needless to say, my bugged out eyes were transfixed on that little rock anchoring the sling into the cavity until I dropped out of its sight. So did the rock stay jammed in the cavity? I wouldn't be writing this story if it hadn't! My palms still get sweaty when I think about my rappel from hell!

NEVER SPIT INTO THE WIND

Wheeler Peak

The drive across Nevada on US 50 was mesmerizing; "The Loneliest Road in America," the road signs proclaimed. I was on my way to the Grand Canyon once again and Great Basin National Park was kind of on the way. At times, you had the feeling the road must be closed ahead there was so little traffic, not a shopping mall or chain drive-in for at least miles! The perfect drive in a roadster some day-long straight stretches interrupted by desert mountain ranges and plenty of curves, over and over again.

I decided to stay in a motel in nearby Ely, Nevada, and just do a couple of day hikes in the Park. The next day, I started up the easy trail to the summit of Wheeler Peak, the highest point in the Park when I met a group of Boy Scouts coming down the trail who said they had given up the hike to the top because the winds were too strong. They advised me to turn back too. I thought for a minute and said thanks for the advice, I'll just go a little further. I was being kind. *How bad can it be?* I thought. I'm in the middle of the Great Basin, it's the first week in June what can happen? The weather forecast said a chance of thunderstorms in the PM, fairly normal for here and I would easily be on top before noon. My ego won out so I pressed on through the gusts that got stronger the higher I went up. The sky began to cloud up and darken but by now I was nearing the summit. A giant gust of wind literally stopped me in my tracks. I distinctly remember sticking my tongue out saying it's gonna take a little more than a wind storm to stop my ego!

Oops! No respect, and so all hell broke loose. By the time I was on the final ridge leading to thirteen thousand footer's summit, well above timberline, the wind was so strong in gusts they pushed me off my feet and knocked me down. It began to snow heavily and soon I couldn't see a thing in the howling wind. I soon realized that I was on the worst place possible for a wind storm—on a ridge high above timberline. Although not cliffy on either side of the ridge, it was as exposed to the wind as you can get—so I dropped fifty feet off the ridge on the lee side on a moderate scree slope and the winds died down enough there so that I could walk again. I could also see alpine lakes far below occasionally when a snow gust let up. Since the trail was along the now impassable ridge I decided to just descend the trailless scree slopes all the way down to the lakes where I knew a trail awaited, and it worked.

When I got back to my motel room that night, the big story on the news was a surprise June snowstorm that had dumped a lot of snow at higher elevations in Nevada and Utah and that hiker rescues were going on all over the place. Just possible showers had been predicted the day earlier but apparently a low pressure system had rapidly intensified. No fun being above timberline in any type of storm—stressful.

A good word for Great Basin National Park—one of my favorite places.

They start you out with a goody in the Ranger Station—a cut out of a Bristlecone Pine that has nearly 4,900 rings on it. The most impressive thing I have ever seen in a Ranger Station—and to think they brag about a cutout of a Douglas Fir that has a mere nine hundred rings on it at Longmire, on Mount Rainier.

Then comes the tour of the caves, formerly known as Lehman's Caves. What a beautiful extra perk in the middle of nowhere. Go on "The Drive" after the cave tour, that takes you up from desert sage brush scrub through the piñons and junipers, up through the Ponderosa pines, Douglas firs, Lodgepoles, Bristlecones, Limber pines and a lot of other trees, right up to alpine meadows with stunning alpine lakes—Stella and Theresa—in them with clumps of Aspens and wildflowers along their shores. Unbelievable variety of plant along a seven-mile drive. From the lakes is the trail through the ancient Bristlecones, their five-hundred-year-old stumps so rot-resistant they only get worn away by snow crystals blowing across them. The ones below timberline only live a few hundred years, the ones at timberline live many thousands, as if the extra hardship was some sort of longevity blessing. Everyone is a different shape created by timberline winds, no two even nearly alike. And one more perk, the only glacier in the Great Basin tucked in on Wheeler Peak's north side and mostly covered in rock debris—-and of course the trail up thirteen thousand feet Wheeler Peak. A day hiking heaven.

SOME MEMORABLE PEAKS

Mount Robson

I was within 750 feet of the summit of the highest peak in the Canadian Rockies when stopped by *verglas*—a thin coating of invisible ice on rock. I was doing just fine until I got to a twenty-foot step that I had to overcome in a weak spot. Not a problem so far, there had been several of them, but this time when I went to use a foothold, it just slipped away. I put my hand on the rock and it was icy. I must have hit the freezing level and the frozen dew from the night before had not yet thawed. I was three days in, so injury was not an option. I could easily have injured an ankle if I continued and slipped on a ten- to twenty-foot rock step, so I very reluctantly turned back.

Mount Edith Cavell

Another especially long seven-thousand-foot climb near the Icefield Parkway in Canada. Failed first time due to showery cloudy weather, second time was successful, class three to four for probably six thousand feet straight. No matter how used to heights you are, the seven-thousand-foot exposure on the summit will make your stomach jump when peering down it. Someone had actually hauled a large cross to the summit in two pieces and erected it.

Mount Assiniboine

I remember getting to the Canadian Alpine hut at the base was just as hard as the next day's climb itself because of ten-foot snow cliffs that had to be overcome to get there. I also remember starting out at the crack of dawn, trying to be the first out of the hut because rock fall was a big problem here and I wanted to be making it, not getting it. I took but never used the rope, just dead weight. One LONG steep face ascent. One two- to ten-foot limestone step after another. When I signed the summit register I realized I was lucky, first one in several years because there was no August snow on the face. No one else from the Hut made it that day, not sure why they were all gone when I got back. I think they took too long setting up belays on what I thought was fairly easy 5.0 steps, most pitches being class three and four limestone. Many entries in the Hut Log detailed weather failures.

Mount Sir Donald

The odd thing about this climb is that as I was registering to climb at the warden station, someone I knew from where I lived came in. He had just been to Sir Donald and failed due to weather. He tried his best to talk me out of my climb because it was raining at the time but I liked the weather forecast for my summit day. He wanted to climb something further east where it would be drier in a rain shadow. I stuck to my guns and he thought I was being foolish, just wasting my time.

Of course the weather did clear the next day and I did summit, but what a long, long fifteen-hour day. I did have to rappel back down on a pitch or two, it was just too steep to down climb. I carried two quarts of water and planned to replenish using snow on the way up as usual, but there was nary a patch. Probably the hardest climb I have done technically solo and maybe effort-wise as well because I eventually ran out of water during the fifteen-hour day and became quite dehydrated. Even though I rehydrated in camp when I got back I still suffered heat exhaustion, just laid there sweating away in the dark on my sleeping bag for an hour until the water I drank caught up with me.

I could see a party still on the summit block coming down a different route than I did. I don't know if they summited but they certainly did not get down before dark. I could see their flashlights and voices—benighted! Probably no water as well. They were fine but caught by darkness and would have to wait until dawn to continue—not a comfortable night for sure but certainly survivable with a coat and hat—and they wouldn't get further dehydrated at night.

Mount Moran

Classic stuff by any standard. Long but flat backpack in, ridge and face climb for many thousands of feet, 5.0 pitches with hand and foot holds as sound as they get. Then glissaded the Frying Pan glacier all the way down, being VERY careful to keep it slow with the ice ax brake, because going too fast risked losing control, not good above a large shrund that you have to exit stage right before you get there in order to reach a forested ridge. An hour to spare of daylight thanks to the time cutting glissade, a 5:30 a.m. to 8:30 p.m., fifteen-hour climb from camp.

Mount Owen

A route finding beauty! Three thousand feet backpack in, no switchbacks. The face was a maze, constantly moving left and right as I went up—and they don't call it the corkscrew route for nothing on the summit knob, followed by a 5.2 chimney, thirty feet ascent between a split summit block—oh yeah more classic stuff!

The Grand Teton

Very few unguided, parties unlike on Rainier.

First time I went with someone I met at the Climber's Ranch who turned out to be more of a rock climber than a mountaineer. He kept belittling me because of all the extra clothing I was taking because the weather forecast was good. However, he forgot about the freezing level. It was in the low thirties near the summit of the thirteen-thousand-footer with a stiff wind and we had to wait at the bottleneck rappel point for an hour while the guided group ahead of us did a long, hanging, 150-foot rappel. All he had basically was a long-sleeved shirt and shorts on while I had a lightweight parka, hat and windbreaker. He shut up about my strategy then when I him use my wind breaker.

Another climb we had to change routes because another rock climbing group from back East had convinced my group that the route I had chosen—the Exum Route—was too hard for them, when I was convinced it was not because I had done it before and knew the capabilities of my group well. The third time a different group of us got into a big argument about proceeding on a perfect climbing day, and nobody summited as a result. The irony was, as we debated, another party showed up, and up they went, commenting to us we bickered that "some things were not meant to be."

Twelve Cascade Volcanoes

Baker, Glacier Peak, Rainier, Adams, St. Helens, Hood, Jefferson, North, Middle and South Sisters, Lassen Peak and Shasta. The Washington ones have incredible glissading as well as Shasta. Rainier is by far the hardest and most dangerous. All are crowded on their dog routes except possibly Jefferson.

All can be done in a day by a strong hiker except Rainier, there you have to be exceptional, only around 2 percent do it in one day.

Okay if I had to pick a favorite it would be Adams, but not by the dog route, although the glissade on the dog route is not be missed at least twice! It's just great camping almost anywhere along its slopes below snow line and dryer because it is further east than the rest of them.

All the volcanoes are mostly long slogs up moderate snow slopes and glaciers that get you in great shape, followed by a glissade descent that takes you down in minutes, what took hours to get up.

Mt Shasta

Mount Stuart

Beautiful granite peak in Eastern Cascades. Long three-thousand-foot couloirs galore, picking the right one means success, wrong one failure. Perfect, wide open, flat camping on Upper Ingalls Creek where most of the many routes up the mountain start after a night's camp.

Mount San Jacinto

Not because of climbing it—there's a trail to the top—but because it is such a huge desert peak. Twenty-five water-less miles to just come down it on the PCT. There's a trail that gains 10,000 feet on the eleven thousand footer. Aspens on top, Mojave desert at the bottom. Only Mount Rainier has that much elevation gain on trail/ route start to top, and only on the Camp Shurmann side.

Mount Whitney

A *loooooong* trail in for a day climb, twenty-plus miles round trip? Came out by flashlight on the trail well after dark. So crowded you could call it an extension of a Los Angeles city park. That same evening drove to Death Valley to sleep in my car at the lowest spot in the lower forty-eight.

Mount Constance and Mount Olympus

Two very different climbs. Olympus is mostly glacier travel after an eighteen-mile approach Constance NE ridge—mixed snow and rock and a very complicated route involving nearly every element of mountaineering, perhaps the finest climb in the Olympics.

YOSEMITE MYSTERY

Once I was in Yosemite and decided to ascend Half Dome via the cable ladder. After a long hike in I was amazed to see "trail" going right up one of Half Dome's cliffs. They had bored holes in the rock, cemented pipes in them, and connected the pipes via cables to use as handholds. Across the pipes at the bottom were wooden planks to stand on. So you pulled your way up the chains from one wooden platform the next, not for the faint at heart or out of shape for sure. In fact I was astonished the NPS had put it there, it looked like a rescue waiting to happen.

So I went up it and back to the starting point. I then decided to go cross country instead of using the trail, in effect doing a two-mile long switchback cut, meeting up with the trail again far down slope.

After going only going only a few hundred feet I came upon day pack rotting away on the ground. It had obviously been there for some time. I checked it out and it was full of things—A back-packing stove, compass, clothing, flashlight—and a camera. Once home I had the film inside the camera developed just out of curiosity and the pictures were of a party that had first gone to San Francisco and Fisherman's wharf. The next pictures were of them ascending the cable route and one last one as they were descending. I wondered for years what had happened, why that pack ended up where it did. Did the owner fall? But how then did the pack end up away from the base of the cables? Then one day an astute woman gave a perfectly logical explanation. You know I'm not going to say what it is, I'm going to see if you can figure it out too! I'll put the explanation on the last page of the book.

GRAND CANYON IMPRESSIONS

I've been asked, "So what is it like to be in the Grand Canyon, to see it from the bottom rather than the top?" A show I saw on television once that featured the Grand Canyon explained it quite well. The reporter interviewed various and sundry people in the depths of the Canyon as he came across them: a woman was crossing the Bright Angel Trail bridge over the Colorado River near Phantom Ranch.

He approached her and asked, "Now that you've seen it from the bottom and are on the way out, what are your thoughts on the place?"

The woman grasped for words for a while and, finally, stuttered, with tears in her eyes, "I never expected this!" Her head shaking back and forth, she threw her hands and eyes toward the heavens and repeated, "I never expected—THIS!"

The reporter approached an elderly gentleman and his grandson who were just finishing up a three-week private rafting trip through the Grand Canyon. The boy, about nine, was weeping silently.

"What's wrong?" said the reporter. "Didn't he enjoy the trip?"

"Quite the contrary," said the grandfather, his voice breaking, "He enjoyed it too much. He doesn't want it to end."

The reporter asked a dory guide his opinion of the Grand Canyon. "You've been down this river many times. What is it about the Canyon that impresses you the most? What do you think is its most spectacular feature?" The guide started to say something but then paused. He appeared to be in deep thought.

He finally said, "You know, it's the way it affects people. They come as quiet, reserved, sort of stand-offish types, but by the end of the trip, they are all a bunch of eight-year-old kids playing at recess."

Yes, that's the Grand Canyon. Your second childhood is waiting for you down there in a sort of bona fide Twilight Zone. I think it is partly the overwhelming sense of isolation. You might as well be on the planet Mars as far as your mind is concerned. The "world" is way up there so very far away, and so very irrelevant to the one you now find yourself in. It's a fresh start. It's a new day, and it's going to be an adventurous one.

On one trip down the Tanner Trail, we made camp halfway down with a most impressive view before us. One woman suffered sensory overload. She perched herself on a ledge, in a sort of trance looking out, and every once in a while she simply muttered, "I'm in the most beautiful place in the world," until darkness took the view away. And you know something? She was exactly right.

My First Time

I'm one of those people who just didn't get it at first. I read somewhere that the mind rebels at having to accept the vastness of the scene you are witnessing for the first time at the Grand Canyon. Well, that certainly was me. On my first trip there, I had gotten to the Grand Canyon late in the afternoon, checked into a lodge, and went over like so many millions before me to take my first glimpse. I have to say I was not overly impressed. I had seen many, many pictures of the Grand Canyon, and the view pretty much jibed with them. I went back to the lodge and prepared for my five-day backpack trip. Bright and early the next morning, I had breakfast and drove to the New Hance Trailhead. It was clear and crisp, my breath easily visible in the early November morning chill. A bank thermometer, as I drove by it, registered nine degrees. Still not overly impressed by the view, I hoisted my pack and started down. I got no further than the second switchback, when I glanced out at the unfolding scene.

My mind suddenly said to itself, "Wait a minute! That's not a small rock I am looking at over there, it's a huge rock!"

The shock of this realization was so profound my jaw dropped and I literally slowly plopped down on my butt, pack still on, and simply stared in astonishment for some time. To add to the near dream-like nature of the place, by the time I was on the river that afternoon, the temperature was in the low eighties, the nine-degree chill of the rim that morning now an incongruent memory of the day.

When I worked in Yellowstone many years ago, we spent many an evening and weekends at the Old Faithful Lodge (the most beautiful building in America!) having a brew or two, and watched many dozens of Old Faithful eruptions. The comments from the kids viewing the spectacle were invariably the same: "That's all, Mom? That's it?" They had built this geyser up in their minds to be one thousand feet tall, and a mere one hundred feet could not possibly live up to their expectations.

Quite the opposite was happening here at the Grand Canyon. Many people, I am sure, were disappointed at their first view like I was, not because it didn't live up to their expectations but because their minds refused to accept that infinite vastness they were observing. It simply so far exceeded their expectations that they could not believe it to be true. I wondered how many other people had stared from the South Rim with uncomprehending eyes, mind in rebellious shock, with a vague sense of confusion about the view, and thereby missed the whole show. If only they had just ventured below the rim to let the reality sink in. When I got back to the rim, I felt like shaking the tourists at the overlooks and saying to them, "You're not REALLY seeing it! You HAVE to go below the rim!" Not out of a sense of "I've accomplished what you haven't so let me brag to you," but more to point out a psychological curiosity that was occurring here.

On my way home, I stopped in Tusayan to watch an IMAX movie called the "Hidden Secrets of the Grand Canyon." I felt so sorry for the nonbackpacking tourists in the audience. Here was visible proof of what it was like down there. The all-encompassing IMAX screen left no doubt as to what they were missing by not venturing below the rim.

For six months, all I could think about was getting back. The following spring, I went again, and it has been an addiction ever since. After many times down, I still view it as an impossible place. Such intricate, infinite vastness cannot possibly exist. Yet there it still is every time I return, defying all imagination and reason. So if you have not been there, I implore you to go while you are still physically able. Anyone who has been to the bottom will agree—not a one will say don't go.

TRUE GRIT

Transfixed by Class 10 Lava Falls Rapid

Legends are not a thing of the past in the Grand Canyon, they are being made today. I stumbled upon one in progress when I signed up to go on a dory trip with Grand Canyon Expeditions.

The juxtaposition between the two worlds abutting Lava Falls rapid could not have been more severe. Look in one direction and you witness the most idyllic Winken-Blinken-and-Nod scene imaginable. You are, after all, in the most scenic spot on Earth, floating in a justifiably glorified rowboat on what appears to be a glassy green pool, surrounded by multi-hued cliffs. But turn around to face that pesky roar in your ear, and watch in amazement as that calm water suddenly sluices down and explodes into the most daunting of frantic, frothy, foamy, stationary wave imaginable in what basically is a horizontal waterfall. Last chance to swim away before you are inexorably drawn into that frenzied maelstrom on a piece of 'driftwood', your outcome anything but certain!

But wait! The driftwood has an oarsman. In my case a very outgoing oarswoman named Amy, one of the famed Hard Core Chicks, so named in great respect by the boating community because of their dedication to and love of the sport of river running in a dory. When not rowing a large rowboat with four other people in it for 275 miles in a 105-degree heat (on a cool day in May, imagine July!) she relaxes with her husband by teaching ultralight flying. You might say she is used to entering the danger zone: she has already been through Lava Falls forty times. The other half of the Hard Core Chicks is Kate, a veteran of 30 times through.

Our lead boat has just gone through safely and it is our turn to take on Lava Falls Rapid, rated a full ten by Lew, our leader, a veteran of fifty times through. As we sink down into a hole below a permanent twelve-foot wave, it becomes quite apparent, yet again, that we are not going to just ride over the top of this one either, and this one is the biggest yet. We are going to plow through it from underneath and hope we are still upright when all is said and done. All three have flipped here before. Touching a single rock, hitting a wave too low, or having the front end stall for any reason ensures instant swamping, flipping and dumping.

The night before I had found a twelve-pack of old stale Keystone beer from 1989 hidden within a Tapeats sandstone cavity near camp. I decided to call it litter, took it back to camp, emptied the brew into the river, crushed and stored the cans. Now I was hoping I had made the karma necessary to get me through Lava intact.

Until you are staring down the biggest wave on perhaps the biggest class ten rapid on the planet, you really don't quite comprehend what "Oh my god!" really means! If THAT doesn't drop your jaw, nothing else will. And just as sure as the sun rises in the east, you are about to go through it. The passengers? True, the guides cared deeply about their welfare. But basically, we were just along for the ride as ballast and to pay for the food. The entire show was in the able hands of the dory guides and their magnificent boats.

By now I know to turn my head so the blast of water isn't shoved up my nose when passing through it. Everything turns into blur of water and motion as we hit. We are heaved and tossed like a rag doll, but several seconds later I see daylight and miraculously I am still in the boat. We made it! Amy hit it just right once again. We turn around and watch Kate's boat go through and cross our fingers.

Guiding four people through class ten Lava Falls in a wooden rowboat has got to be the gutsiest thing I have ever seen done. (Watching a Havasupai Indian catch a wild horse without using a horse comes in second.) Even old John Wesley himself didn't have the grit to take on the Biggies. They used ropes to lower their craft through some of the rapids. But then they were doing it for the first time. Mere pikers compared to our gang. I will never forget going through Lava Falls. It is right up there with glissading down a vertical mile from the top of Steamboat Prow to Glacier Basin on Mount Rainier. Joy-wise that is. For sheer excitement nothing else I have ever done outdoors compares. (Well, except for maybe dropping 2,500 feet in an avalanche on the Brothers). That the Park Service even allows such undertakings amazes me. Already that spring, someone had drowned in Crystal Rapids before our trip, and we witnessed two rafts flip ourselves on comparatively minor rapids, helping one of them to unflip.

After Kate gets through we pull up at the nearest beach and give thanks to the river for safe passage. A bottle of Tequila is passed around to a grinning bunch, and Kate makes the offering while pouring some into the Colorado. Other than the true grit of the guides, the perception that sticks with me was the sheer joy of going through the rapids. I never would have expected it before the trip. I have a slight phobia of water and would never venture out into placid Puget Sound in a rowboat with a wind of ten knots. And I am not a great fan of cold water, refusing to swim in water less than sixty-five degrees. So here I was, being taken on a fourteen-day trip through the biggest rapids on the planet and in water not much warmer than an alpine lake. I anticipated having to suppress fear and apprehension and having a bit of misery from being coldly wet. Sheer joy was not on the list of expected emotions.

But there is just something about being in the middle of a large rapid that is unique. I think it's a combination of sensory overload, with a refreshing yet all-encompassing roar so loud it is a constant presence in your mind, along with the vision of a world compressed to and composed entirely of frothy bubbly white water. It seemed preposterously beautiful and exhilarating. And all the while you are pitching to and fro in every direction imaginable. You are keenly aware that the slightest misjudgment by your guide will send you tumbling into that gargantuan Maytag rinse cycle, your head certain to be under water for seconds, life jacket or not.

And yet you feel sheer joy. The looks caught on pictures of people's faces as they passed through the rapids confirms this. I vividly remember giggling uncontrollably. Was this some sort of addiction for the guides, the feeling being so pleasurable? The looks on their faces in the pictures confirmed it when I zoomed in. They lived for these moments. Kate (as well as Amy) was the epitome of a young woman at her physically fit best think Olympic competitor tough. Her other love (other than her husband) seemed to be anthropology. She was a treasure trove of info on the people that used to live here. I finally got the big picture of what was basically a land rush by the Ancient Ones who colonized many, many dozens of side canyons at the same time, perhaps ten thousand people in all. Then a huge drought ensued and while they still had water, others in the area did not, so out of desperation they raided the riverside civilization out of existence. We stopped along the way to witness an archeological dig in progress, allowed by the Indians only because erosion was in danger of washing all artifacts away. A vast area of the Canyon has yet to be fully explored, archeologically speaking.

Lew has a twin brother who runs a ranch in the area. Lew decided instead to make the Colorado River corridor his ranch, the dories his bucking broncos, and we the clients on his dude "ranch." An easy going fifty-year-old with an affinity for telling stories, especially on Grand Canyon history, Lew has no plans to retire soon, but he has a long way to go to catch his mentor and former boss and dory pioneer Martin Litton, who piloted a dory through Lava at age eighty-seven. Like Mr. Litton, who was instrumental in leading the fight against damming the Canyon, Lew is well versed in current Canyon controversies.

The wandering souls of both Jim Bridger and Sacagawea had drifted here to find a new niche in twenty-first century America in the forms of Allen and Katie, our supply boat crew. Allen, a large, kind of gruff guy, was the official photographer and a jack-of-all trades with a heart of gold.

Katie, a twenty-four-year-old Blackfoot Indian with a degree in ecology, was Allen's swamper, river talk for free help in exchange for the ride.

One of the surprises of the trip was that Allen, a veteran dory guide himself, gave the supply raft rudder to Katie as they neared Lava Falls and let her pilot the supply raft through with his guidance, a MUCH, much easier task than the dory guides had because the supply raft is so large and has a motor. Presumably, she had practiced on a few lesser previous rapids. What an experience gained and bragging rights for a newbie. In the end, Katie was so well liked that the crew gave her an equal share in their earnings. A future Hard Core Chick in the making?

Another treat besides the obvious physical ease (for the paying customers) of doing this trip, was being able to explore the riparian environment of the Canyon each and every day, instead of only occasionally after a grueling hike down dampens enthusiasm for exploration. Amy and Katie were botany enthusiasts and we had a couple of geology professors among our total population of 15. I learned many new plants and a few new geology tidbits. The most common animal we saw was the bighorn sheep and there were plenty of them. They must inhabit every side canyon. Of course, the ravens were always there to clean things up, and Sacagawea, I mean Katie, almost stepped on a large Grand Canyon Pink Rattlesnake. I pointed it out to her after she had passed almost right over it, and I was about to. I hope she was impressed by my luck—I mean my outdoor wisdom!

The side hikes were many but short, to accommodate the aging, out of shape—but well-to-do—clientele. Besides, it was 105 every day, and the crew was getting plenty of exercise already from rowing in the heat. Other than myself, the only real hiker in the paying group was Bill, seventy-four, who had hiked the entire John Muir Trail the previous summer. His son gave him this trip as a birthday present. An immigrant from Germany as a teen, he had vivid stories of living in Germany as a boy while it was being bombed to hell and then invaded by the Allies.

The cold water turned out to be a blessing. A quick submersion and you were cool again. Plus, it kept the beer, shrimp and steaks cool and fresh. A rotating staff of guides turned chef came up with a winner dinner worthy of any good restaurant each and every of the thirteen nights.

All and all a mighty fine trip that I would not hesitate to do again.

And what if I had just left that beer can litter that I then would have known was there? I am extremely superstitious about litter. Would I have been tossed into Lava? We will never know, now will we, but perhaps knowing is not the point—and not knowing is.

JOYCE'S RESCUE

I had organized and planned a trip to the Grand Canyon with a bunch of fellow hikers belonging to the Peninsula Wilderness Club of Bremerton, Washington. We were on the Tanner trail and had decided to make a water-less camp on a plateau above the Redwall formation, about halfway down. This is possibly the best camp for views in the entire Canyon. You get reds from formations not found elsewhere in the Park. Sally, one of the members of our group, had become entirely mesmerized by the scene. She perched on a rock and every few minutes she shook her head and exclaimed the same thing, "I'm in the most beautiful place in the world."

Hard to argue with that.

The next morning we broke camp and began the second day's descent to our next camp on the Colorado River. One woman, Joyce, had brought a pack that must have weighed fifty pounds, with all kinds of clothing that she would not need for a hot May day in the Canyon. She wore heavy "broken in" mountaineering boots, but there was a slight problem. The upper eyelet had broken on one of them, so the boot could not properly be laced tightly around her ankle. It was quite loose instead. As she started around the second switchback out of camp, she slipped on some gravel-like rock that was on the trail. She tried to catch herself with the loose boot and POP went her ankle. Down she went. After wincing in pain awhile she arose and said she was fine. She tried walking uphill. There was only a slight limp. But when she turned and tried to walk downhill, she screamed out in pain and collapsed yet again. "It just needs some rest," she said, not wanting to face the obvious.

After five minutes, she tried again to walk down hill to no avail. So now I was faced with a tough decision: Call off the entire hike, ruining a long planned trip for eight people, or not. Joyce kept insisting we go on, not wanting to ruin everyone else's trip. I quickly decided that there was no way that Joyce could continue. I realized we were very near to last night's camp on a very flat mesa where a helicopter rescue was quite feasible. The lower we went the worse off Joyce would become, and the rescue would become much harder. They probably would have had to hand carry her out on a stretcher up some three thousand feet. In addition, there was no water source anywhere near where we were and we only had enough water each to make it to the next night's camp on the river. Joyce simply could not stay long where we were.

"You stay here with Joyce, Sally, since you have nursing experience. The rest of you continue on down to the river after leaving a half quart each of water for Joyce and Sally. I'll hike out and get help."

I left most of my gear, taking only enough water to make it to the rim—about half a quart. The hike out went quite fast and I was calling 911 from nearby Desert View visitor center by early afternoon. The ranger there was skeptical. "They don't usually helicopter people out with sprained ankles, he said. "They usually have more important things going on." But Joyce was at least somewhat lucky that day. There were no life threatening rescues going on so they sent a helicopter right away. Thanks to my explicit directions, I watched as the helicopter descended and immediately went to the flat mesa and found Joyce within fifteen minutes of my call.

Joyce later said the very first thing the rescue guys said to her after hopping out of the helicopter was, "I know exactly what happened! You were walking down a steep trail slipped and tried to catch yourself and bang! Twisted ankle."

"How did you know?" Joyce said.

"Second most common injury in the Canyon," the rescuer said.

Sally said, "I have to ask, what's number one?"

The rescuer said, "Rattlesnake bite between the thumb and forefinger. Dudes try to impress their girlfriends when they come upon them by showing they can grab them before they can strike."

I drove over to Grand Canyon Village to await the helicopter with Joyce in it. I couldn't help notice all the pictures they had plastered on the wall showing boating accidents. In one an overturning raft was just about to land on top of someone who had fallen out of it, his hands held up in an effort to stop it.

Turns out the helicopter went to check out a report of an overturned raft on the Colorado. It turned out to be nothing but Joyce got a rare treat—a flight by helicopter into the inner gorge where no recreational planes are allowed to fly. She got to experience something few tourists get to see by air. After avoiding a thunderstorm for ten minutes, they finally landed and Joyce's ankle was put in a brace. After a night in one of the lodges, Joyce decided to stay at a campground on the rim while we finished our hike, and make a leisurely tourist thing out of it. Temps dipped into the midtwenties at night, she later said.

I went back down the Tanner, picked up some water Sally had left and met the group on the sandy banks of the Colorado as planned and we finished the hike without further incident.

Someone predicted Joyce would get a big bill for the helicopter rescue ride and treatment at the clinic but her health insurance covered it all saying it was an ambulance service.

WHY YOU SHOULD NEVER EVER LITTER!

Once upon a time, I was helping the Olympic College mountaineering class on a rock climb of Guye Peak. On the way back down, we set up a rappel point. These places usually turn into bottlenecks for groups so most of the students kicked back and awaited their turn to rappel. One woman was a smoker and as she finished her cigarette, she crushed the butt out, left it on a rock, stood up and got ready to descend. About ten feet into her rappel, she leaned too far forward toward the cliff that her feet were on. As she neared a near standing vertical position, instead of the desired forty-five-degree angle, her feet slipped out from under her. Before she had time to react, her head was shoved forward by her weight into the cliff, splitting her lip. Instant karma! Luckily she didn't let go of the rope.

A long, long time ago, in an area not so far away, I was on a climb of Mount Stuart. We had lunch on the summit and one guy had brought a cantaloupe. Despite our objections, he heaved the rinds off the summit, sneering that they were biodegradable. On the way down, we had a tricky move to enter a steep gully that required the use of handholds. We all passed safely, testing the handholds as we went. Except, that is, for the rind tosser. As he entered the couloir, he neglected to test his handhold. It immediately came loose, he lost his balance, and we watched him fall ten feet, doing a complete somersault in the process. Luckily, he landed on his pack, but that didn't keep him from developing a knot the size of a football on one of his bruised thighs. Almost instant karma.

As I was preparing at the Tanner Trailhead to descend into the Grand Canyon, a hiker came out looking for help for his wife who had twisted her ankle about a mile down the trail, ruining the start of a multiday trip for them. They were taking a break under a shady overhang, got up to go on, and she simply slipped and fell on the trail. As he lit up a cigarette, he went to his car and said he was going for help, and could I check on his wife as I went down. I said sure. About a half mile in, I came upon her with four other hikers helping her to hike out using a pair of walking sticks and their support. Since the trailhead was so close and I knew more help was on the way, I decided to continue on down after they said my help wasn't really needed. About another half mile in, I came upon the obvious, shady, and only overhang around, where the pair had taken a break and, lo and behold, there were two cigarette butts crushed out on a rock. Uh hmm. I rest my case!

PARIA CANYON FLOOD

This is a story I have meant to write for some time but have been reluctant to do so. Why? I have had a number of close calls in my many hundreds of mountaineering outings (only one injury, though and that was as a student in a class). Nearly all were due to objective hazards such as rockfall, AM thunderstorms, etc., that could only be avoided by not going at all, rather than poor judgment. But Paria was the only one where, although not directly my fault, the entire party was endangered.

I had first heard of Paria Canyon on a Grand Canyon hike. I was in the middle of nowhere as usual when a fellow hiker passed by. He was an experienced desert rat and highly recommended Paria Canyon. Now I am somewhat obsessive/ compulsive in nature. Sometimes I see a peak on a calendar or check book and that is it: my next goal. I immediately decided that Paria would be my very next desert hike.

I bought the guide book and trail map and set about organizing the trip, offering it up as a club outing. To my surprise, twenty people signed up. Since the limit for the hike was ten, I divided the group in half, with one party to go upstream and the other down. I secretly decided which group to join, the upstream one, because I felt they were a more dedicated group, more likely to complete the hike. This is not always a good thing as you will see. I am by no means a disciplinarian, so I let people decide on their own what group to join, gear to take, etc., because all were experienced backpackers.

Paria is a slot canyon of monumental proportion so I carefully chose the best time of the year to hike, late March, when the likelihood of flash floods was at a minimum according to the guide book. The travel portion of the trip went like clockwork with all twenty people arriving in Las Vegas on several different flights from several different cities. We all rendezvoused at McCarran airport, got our rental vehicles, and off to Kanab, Utah we went for a night in a local motel. We strategized how we would each travel to different trailheads and how we would return separately to Kanab after the hike. At first we decided to swap vehicle keys when we met in the middle of the hike but abandoned that idea in case something happened and for some reason we didn't meet.

We decided to hide the keys somewhere on each vehicle instead. I traveled with my group to Lee's Ferry on the Colorado River to begin our upstream hike. Now the weather reports on the TV had varied. The local NWS predicted a minor front to go through midweek with scattered showers followed by clearing, but the Weather Channel was predicting rain by midweek. This was talk of some concern given the nature our hike, so as a final precaution, we checked into the ranger station at Lee's Ferry and, after some consultation via phone, the ranger gave us a thumb's up, basically agreeing with the NWS. So off we went.

The first two days of the hike were pretty atypical of desert hiking. The weather was in the 80s, people were getting sunburned where they forgot to put sunscreen, but all and all a pleasant stroll up the twenty-foot wide ankle deep "river'. The main difference was that the river had to be waded numerous times due to the cliffy terrain, so wet feet were the rule. We all carried extra footwear and socks for in camp use. The canyon was by no means narrow at first, probably a quarter mile wide those first two days. The water was quite silty, causing water filter failures nearly immediately. We had a very generous schedule for the forty-five-mile, seven-day hike, only about seven to eight miles a day with very little elevation gain. It worked out to about five hours of hiking a day.

The third day I had scheduled as a rest and exploration time. There was a large arch near our camp and several of us went to have a look. On the way back there was a dramatic, and I do mean drastic change of weather. An enormous wind/rain storm came in within minutes. Luckily I was near an overhang and was able to stay dry. But as I stood there I could see several members of the party hiking on a hillside across the river. It was obvious they were getting soaked, with the wind blowing so hard that they literally could not get their rain gear on. After I got mine on in the shelter of the overhang, I hiked the remaining distance to camp to find tents with wind-detached flies, their contents soaked. Several people had not even pitched tents, and their gear was soaked as well. They had hiked in the desert before with little or no rain so they had opted for minimum shelter, a very common strategy for desert hiking. I met another party member in camp.

Although campfires were banned in Paria wilderness, the first thing I said to him was, "We are going to need a campfire to dry things out."

His only response was, "I'll start gathering firewood."

Although I had no tent, my gear was dry. I had picked a slight cave at the base of an overhang as my camp spot. In addition there was a large rock just in front of the mini cave that kept windblown rain from entering my camp. We built the campfire under another overhang using a large dead cottonwood tree for fuel and people spent that evening drying out gear and making new camps out of the weather. In hindsight, this should have been done before the monstrous storm not after, but, hey, we all make mistakes. There was of course concern of a flood, but we were not yet in the narrow portion of the river so escape from one was not really an issue. It continued to rain, and rained all night but at least everyone was now reasonably dry.

At first light I awoke. I could see up the hill on the other side of the river and was astonished to see that it was white. Snow had fallen a mere thousand feet uphill. The Paria had risen at least a foot to knee deep and was the color of chocolate milk. I was surprised that it had not risen further since quite a bit of rain had fallen. Apparently, all the precipitation at higher elevations had been snow. Then as I watched in amazement, the rain turned to snow. At only about three-thousand-foot elevation (the bottom of the Grand Canyon is only slightly less), this was indeed a rare event. I got up and restarted the campfire. People slowly gathered around me and a debate began: What should we do. Some people figured the weather was about to break, so they wanted to break camp and go on. Others wanted to cancel the trip and go back. I decided to compromise and just stay put, reasoning that we were warm and dry where we were, but would become quickly soaked if we hiked in either direction. It snowed all day, eventually piling up four to six inches deep. We stayed under the overhang warm and dry thanks to our exquisite illegal campfire. When we retired for the night it was still snowing.

The next morning broke partly cloudy and blustery with the temp around forty degrees. The debate continued. We reasoned that the storm was over and that we had seen the worst possible weather and survived. I asked if anybody wanted to turn back. To my amazement, no one said yes. So on we went through the quickly melting snow. The canyon now narrowed rapidly. We camped that night where the down-stream group had waited out the storm. They left a note saying they had decided to turn back. The next day the canyon narrowed dramatically, in places only ten feet wide and 1,500 feet high. We could see logs wedged some forty feet up attesting to the depth of previous flash floods. The Paria had risen to waist depth in places and crossings became quite numerous. We were now hiking almost as much in water as out of it. After setting up our final camp, we spent the afternoon exploring Buckskin Gulch, another slot canyon so narrow and so deep that at times it was nearly dark. We did not know it at the time, but it had snowed thirty inches three thousand feet above us. If it had rained instead of snowed, we would have been in big trouble. That night a tremendous thunderstorm hit and it poured for hours. Once again the snow above saved us, absorbing much of the rainfall like a sponge, preventing a flash flood. The Paria had now risen to nearly chest height in places, but the current was still very slow due to the extremely flat nature of the canyon. So here we were with no choice but to go on. Retreat was now impossible due to the length back.

Our last day was like no hike I have ever done and have no desire to ever repeat. We were forced to slosh through water for three miles. The air temperature was in the low forties with a stiff breeze blowing. There was slush flowing down the river along with much debris. I could feel my blood being refrigerated as it flowed through my legs, which became completely numb from the cold. I remember thinking that I wasn't going to make it as my legs became increasingly difficult to move. After hiking for one and a half miles without once being out of water ranging from knee to nearly chest deep, we came upon a sliver of a beach, and miracle of miracles the sun was shining on a wind sheltered cliff, warming it like an iron. We all stripped as much as modesty would allow and clung to the cliff allowing its warmth and that of sunshine to seep into our bodies for quite some time until the feeling—and aching pain—returned to our legs.

Getting back into the water again was not a pleasant thing to do, but we had no other choice. The legs quickly became numb again. If anyone had stumbled and gotten their upper torso wet, there is no doubt in my mind that they would have died of exposure. It was quite literally impossible to get out of the water, let alone having any hope of drying out even if you could. We were all a slip away from disaster. After another mile and a half without once leaving the water we finally exited the slot canyon, and two hours later we were in the van on our way back to Kanab where our worried companions awaited our return.

Several months later I got a letter from one of the young women who was in our group. She thanked me for allowing her to participate in what she considered to be a great, memorable adventure—it was her first backpack with her experienced mother—and finished her letter by saying, in retrospect, that she might now even admit that it was fun. Well, that's one positive way of looking at it! Others weren't quite as generous in THEIR comments.

Several months later a group of twelve was caught in a flash flood in a slot canyon only forty miles east of Paria. They had been warned by the rangers not to attempt the hike, August being the most likely time for flash floods. Their paid guide ignored the warning. He was the only survivor, having clung high up on a rock as a wall of water swept away his paid clients into a virtual concrete mixer full of large boulders. Other people watched helplessly in horror from above.

Moral of the story: The rangers don't control the weather—but apparently, the Weather Channel does!

PACIFIC CREST TRAIL INTRO

CERTIFICATE OF COMPLETION

For having hiked the entire length of the

PACIFIC CREST NATIONAL SCENIC TRAIL

May 3, 1999 through October 30, 1999

Awarded to

David Cossa

Acknowledged and Presented by the

Pacific Crest Trail Association

October 2000

Robert S. Ballou, Executive Director

David Foscue, President

This is to be the story of my six-month journey along the Pacific Crest Trail. I hope to go easy on the superlatives and concentrate instead on events, feelings, and the people I met along the way. It was awesome, incredible, beautiful, magnificent, jaw dropping, butt kicking, and diverse beyond belief. There! That should be enough adjectives for the entire story.

So what is the Pacific Crest Trail? It is one of eight nationally designated scenic trails.

It was so designated in 1968 and finally dedicated in 1993. The PCT runs in a continuous path some 2,650 miles long, from the Mexican border to the Canadian border, through the States of California (1,650 miles), Oregon (five hundred miles), and Washington (five hundred miles). The trail passes through twenty-two wilderness areas and six National Parks, exposing the through hiker to a bewildering array of flora, fauna, and scenic wonders. The chaparral, desert, and various and sundry mountain forests are the major components of ecosystems found along the trail.

The Appalachian Trail, the Nation's first designated scenic trail, was the model for the PCT Various hiking and equestrian clubs, working in conjunction with the US Forest Service were instrumental with their volunteer labor in constructing the Trail. Several sections pre-dated the PCT, including Washington's Cascade Crest Trail, Oregon's Skyline Trail, and California's John Muir Trail. The idea was to link these trails together with new sections. Passing through State and Federal land posed no problems. However, some new sections had to pass through private lands. Surprisingly, most private land owners thought the PCT was a great idea, readily agreed to right of way passage, and even helped with the construction.

Some private land owners, in particular the massive Tejon Ranch in Southern California, were quite reluctant and it took decades to finally persuade them to cede egress through their holdings (one may now hike through the Tejon Ranch, but stepping off the PCT for any reason is considered trespassing. They fear range fires on their arid property). Several other short sections had to be routed around mining ventures leased in wilderness areas. The leasers could not be persuaded to grant egress for fear of theft from ore rustlers.

The US Forest Service now has the overall responsibility for the PCT. Operation is shared by the Forest Service, National Park Service, Bureau of Land Management, and the Pacific Crest Trail Association. The PCTA, through a Memorandum of Understanding, helps maintain the Trail, promotes its use, and advocates for its protection. They organize and conduct numerous work projects on the Trail each year.

Most people hike the PCT in short sections and those that complete it usually do so over a period of several years. So why did I decide to do it over a period of 6 months? Several reasons seem obvious. Being an avid mountaineer, I liked the challenge. Being fifty-two years old, it was sort of an imperative physical mid-life crisis—better do it before it's too late. Another obvious reason was that I love backpacking, and being outdoors witnessing Nature's beauty for six months straight greatly appealed to me. There are other less obvious reasons as well.

Someone perceptively remarked that it was a "Forrest Gump" thing. Like Mr. Gump, I had been rejected by someone I fell in love with and needed to do something to break the negative spell of self-pity that I found myself in. Forrest liked to run so he ran. I liked to backpack so I backpacked like I never backpacked before.

It was also a Thoreau "thing," an emotional component of my mid-life crisis. I wanted to separate myself from the culture I lived in, to become an observer rather than a participant. I had hoped my journey would cause me to examine my values, wants, and beliefs. I wanted it to be a lesson in what to throw away and what to keep. Thoreau's idea of a long separation seemed to fit the bill.

Since one of Thoreau's mantras was to simplify, simplify, I quickly rejected the strategy of meticulously planning, at home, each meal, each day's agenda and each night's camp, opting instead to be spontaneous and to plan the trip week by week on the Trail. I read several PCT books (including one I highly recommend, Journey on the Crest by Cindy Ross), and lightly perused the PCT guidebooks.

I obtained a through permit from the Pacific Crest Trail Association, good for the entire trip. I also purchased a town guidebook from the PCTA listing services such as post offices, groceries, and lodging in towns near the PCT My resupply strategy was to simply buy food at the larger groceries along the way and to mail food parcels purchased at these groceries to towns with limited supplies. This strategy worked very well.

I found a house sitter and a phone contact for trail emergencies, gear replacement, and mail forwarding. For financing, I remortgaged my house and carried a debit and several credit cards.

I jettisoned all my wool and cotton clothing for lighter quicker drying fleece and Supplex nylon duds. I wore Red Wing hiking boots, long my favorite for durability and comfort. I tossed my gas stove for a wood burning one, eliminating the need to find and ration gas fuel. I carried no water filter, opting instead to continue my forty-five-year tradition of drinking directly from streams and lakes (I had no problems).

I worried at first about being lonely, since I was hiking solo. But once I immersed myself in the Journey, I came to love being a nomad not responsible for or dependent on others. This doesn't mean I didn't enjoy the company of other hikers. I met dozens of other cresters. Often we would hike together for a few days or stay together in the way station towns.

To be honest, I expected adjectives like brash, egotistical, know it all, competitive, and braggart to describe the people I would meet. Not one of these adjectives applied to any of them. Instead, I found them all to be confident, reserved, kind, caring, helpful, outgoing, and infused with a live and let live attitude. I even adopted a couple of them as my surrogate kids and found myself hoping for their success more than my own, eagerly looking for their names in registers along the way.

One bonded instantly with other cresters. After all, we were sharing the same dream and the same great adventure. It was easy to spot other cresters. We stood out because of our deep tans gleaned through months of being continuously fried by the sun. Although most of us hiked solo, and sometimes weeks and a thousand miles would pass between meetings, the bond was always there.

There was also the mysterious force known as Trail Magic, and the presence of people we cresters affectionately called Trail Angels. Trail Magic began even before I started hiking. I had planned to drive a rental car to San Diego and take a bus from there to the trailhead. At a potluck dinner a week prior to leaving, I talked of my plans and discovered that someone was driving to San Diego for a temporary job at a shipyard there—at government expense. Think of the odds against that! Trail Magic, indeed! As for Trail Angels, well I'm getting ahead of myself. I've divided my story into five sections and the Trail Angels will get their proper due in due time.

PACIFIC CREST TRAIL PART 1: SOUTHERN CALIFORNIA

Mt San Jacinto

So here I am, at Campo Ca., near the Mexican border, ready to begin a 2,650-mile sojourn to Canada. I hike the confusing one and a half miles to view the PCT Mexican Monument and on the return to Campo someone unseen yells, "Are you a PCT hiker going to Canada?"

"Yes, I am," I reply. "Well, good luck to you. I hope you make it."

It is to be a mantra I will hear dozens of times over the next six months: "Good luck to you." It lightens my spirit and load to know complete strangers are wishing me well.

Campo resembles a police state with Border Patrol jeeps and helicopters cruising the area. A real cat and mouse game is going on and as I learn on my first night out, the mice are winning. A group of twenty illegal aliens pass by my camp. Each of us is relieved to find that we have no interest in the other.

The trail starts out in chaparral, something foreign to me, and I eagerly investigate the stuff. Rigid branches make it impenetrable off trail without a machete. Fifteen-foot-high manzanita, scrub oak, mountain mahogany, sumac, chamis and spiny ceanothus predominate. The intelligent plants are extremely flammable, for they know fire is their friend. They sprout back immediately after a fire. Not so for other intrusive plants, like pines, that would crowd out the chap.

The first mountains along the PCT are the Lagunas, a brief twenty-mile respite from the chap. Black oak, Jeffrey and piñon pines abound. I pick up a self-mailed parcel at the post office in Mount Laguna and stay overnight in a PCT discounted cabin, the first of many specials for PCT hikers.

The next day it's back to the chap after some good views of the Colorado Desert, five thousand feet down to the east. The chap will prevail for the next five hundred miles except for three fifty-mile excursions into the San Jacinto, San Bernardino, and San Gabriel mountains. I coin the phrase 'wrap the chap' because the trail crosses innumerable ravines not by dropping into them, but by contouring around them, over and over again. Occasionally, groves of massive Live oaks, some 8 feet in diameter appear. Well-spaced springs, shared with cattle, are the only water sources.

Warner Springs is the next way station, with little to offer us cresters except a post office overwhelmed with thru hiker parcels, including mine. From here, a 5,500-foot ascent into the San Jacintos starts. Midway through them I drop 2,500 feet into Idyllwild, a busy vacation resort. A free stay at a state park and a resupply at the local grocery are in order. Being a mountaineer, I feel compelled to ascend 11,500-foot San Jacinto Peak, the scenic highlight of southern California. San Jacinto is a massive peak, with a ten-thousand-foot, twenty-five-mile waterless descent into the Mojave Desert.

Crossing I-10 is a bizarre experience. The Interstate is crammed with cars and trucks, completely incongruent with the empty surroundings. It's like a huge conveyor belt, ferrying souls and materials into some great, giant, smoky machine just past the distant horizon. A surrealistic wind farm is nearby, the whirling, squeaky blades slicing up the silence that preceded them. It is a relief to bid them farewell and climb six thousand feet up into the forests of the San Bernardinos. Piñon and coulter pines, with their three- to five-pound cones, give way to enormous specimens of sugar pines, Jefferies and Foxtail Pines, in my opinion the most picturesque of all trees. They are like snowflakes: no two are alike. These in turn yield to ubiquitous lodge poles.

Big Bear City is a very friendly place. The fire station lets us stay on their back lawn for free. The guy at the post office delivers a mismailed parcel to Tim, a fellow crester, at his motel room, after hours, on a bicycle.

Now it's back out of the forest for a ten-mile sampling of the Mojave before ascending six thousand feet once again on this yo-yo path into the San Gabriel's. Water remains plenty scarce with no creeks to speak of. Way station Wrightwood is known as the black hole of the PCT because so many cresters drop out here. The exhaustive ups and downs, and the monotony of the chaparral are more than some can take. Tim is a victim. A vegetarian, he is but a skeletal remnant of his former self and too tired to continue.

A giant storm is predicted, and it comes true. One day out from Wrightwood, a rainstorm that would do the Northwest proud develops and turns to snow. The temperature drops into the low twenties. It's June 1! Passing by a Boy Scout cabin with a wood stove is just too tempting. I enter through an unsecured window, start a fire, get dried out and have a warm, comfortable night.

On the descent to Agua Dulce, the temperature soars into the nineties. I almost walk into a rattlesnake perched in a bush, about waist high. I recoil like a hand touching a hot wood stove. Shuddering, I press on, keeping a wary eye out for more of them.

I take a side trip to a grocery only to find it closed. As I leave, a man comes to the door inquiring if I'm a PCT hiker. When I say yes, he opens the store just for me. I'm becoming dumbfounded at the helpfulness of people along the trail. It's not what I expected.

At Agua Dulce, a woman is waiting at the post office. It's Donna, one of the arch angels of the PCT. She insists that every passing crester stay at her guest house. She does our laundry and chauffeurs us to town for supplies and a feast at the local Mexican restaurant. She makes an indelible impression in every thru hikers' mind for her overwhelming generosity. But there is more to come!

The Mojave Desert is next with its yuccas, blooming cacti, ocatillo, and Joshua trees. It's mighty hot, in the low hundreds. I can only hike about an hour at a time before diving into the scant shade of a Joshua tree. We are following roads along the concrete encased California aqueduct, a sixty-mile waterless stretch except for one small spigot tapping the aqueduct put in for, you guessed it, PCTers. Normally, you'd have to carry about two gallons of water to make it through. But trail angel Donna and her helpers have stashed water for us at prearranged spots along the trail. In addition, trail angels cruise the aqueduct road, looking for thirsty cresters, dispensing Dorito's, orange juice, and other welcomed goodies.

Now comes the ugly part. The trail passes through the Tehachapies. Due to easy access dirt bikers have invaded, or should I say infected, the entire range. Two-thousand-foot bike ruts scar most of the mountains. The drunken sots have littered everywhere, and shot up every conceivable post, sign, and structure. The aqueduct had to be placed underground through here so that these bozos wouldn't destroy it with armor piercing bullets. I come to despise these people for their uncaring attitude toward the land, especially after meeting so many hikers doing their best to go with a no-trace attitude.

So it comes as no surprise that the Tejon Ranch doesn't really want anyone on their land. They let cresters pass through, but only with the caveat that we cannot camp or leave the trail for any reason. It's patrolled on horseback.

When I reach the highway to the town of Mojave, the next supply depot, a truck driver offers me an unsolicited ten-mile ride to a welcomed air-conditioned motel. It's 110 degrees outside. After a two-night stay, the eighty-five-year-old owner, who's been at the motel since 1928, gives me a free ride back to the trail head. I'm refreshed and ready for the Sierras. Six hundred miles down and two thousand to go!

PACIFIC CREST TRAIL PART 2: THE SIERRAS

Well, so far so good. My feet have a couple of small blisters, but nothing memorable. That's about it physically. Not bad for fifty-two! I've been averaging sixteen to eighteen miles a day, slow by PCT standards, but far more than I dreamed possible only two years ago. I'm about four days behind schedule, and expect to slip even further in the Sierras, but I hope to make up time in Northern California and Oregon. People have been passing me almost daily, but as long as I'm nearly on my own schedule (one hundred miles a week) I don't much care. Stay the course I keep telling myself.

It's a long slow climb through the Sierra foothills. The chaparral is finally gone, and piñon pines have captured the trail. So have the bugs! Seven hundred miles of nary a bug before I passed through some invisible barrier that was keeping them in. A long sleeve shirt and pants make all the difference in the world. Without them it would be miserable. With them, the bugs are actually tolerable. Just a little DEET now and then on the face keeps the buggers at bay. Water is now plentiful too. It's a relief not to carry water anymore after carrying three to eight quarts of it every day for weeks on end.

I've gotten pretty good at cooking on the Sierra Zip wood burning stove. Even in the desert there was always something to burn as fuel (it only needs twigs). I'm totally sold on this neat little gadget that only requires one AA battery about every five days to power the fan. I can cook things like potatoes without having to worry about rationing fuel.

Kennedy Meadows is one the PCT's greatest way stations. There's no phone or power here, but the general store has its own generator to provide us with hot showers and a laundromat. There's no post office either, so the owner holds parcels for us in a storeroom. When I arrive, about fifteen Cresters are gathered on the store deck, repacking goodies received in the mail. Among them is Damien, thirty-fiveish, who owns a winery in Northern California and Florian, a Chinese woman from Sydney, in her early twenties. Robert, a very opinionated young black student from Dayton, Ohio has just left.

"Let it be." An unabashed partaker of the weed is about a day back. He has two triple Crowns to his credit, having hiked the PCT, Appalachian, and Continental Divide trails twice, and is working on his third. He is addicted to long-distance hiking and has gone on trips for twelve years in a row. All these people are remarkably easy to get along with, even charismatic in nature, and I will be hiking on and off with them all the way to the Oregon border.

That evening the owner of the Growly Bear restaurant makes his daily trip up to the general store to give hikers a ride to his establishment, three miles down the road. Tonight it's fried chicken and mashed potatoes, and about twenty of us are savoring the rarity of a home-cooked meal. After pie ala mode, the owner takes us back to our camps.

Now the great climb begins. First up, the Golden Trout Wilderness. Piñons quickly cede to sugar pines and Jeffrey pines. California Junipers and Lodge poles are next. Then come the timberline Foxtail pines. I notice something very odd about the "soil" they are growing in: there is no humus whatsoever, just sand and gravel. One gets the distinct feeling that the downed corpses of thousand-year-old trees were the first to grow here since the glaciers ceded their reign.

At first, the mountains are quite rounded, but as we approach Mount Whitney and the John Muir Wilderness their character changes remarkably. Lakes begin to fill every view. One can't help feeling awed as they pass through thirteen-thousand-foot Forester Pass and look north to a seemingly endless array of jagged peaks.

I skip Mount Whitney since I have already climbed it twice and am behind schedule, bailing out instead at Kearsarge Pass. It's a

2,500-foot descent to a road that leads to Independence. The resupply goes like clockwork: a ride in thirty minutes, in the motel by 1:00 p.m., lunch by two. Groceries are bought and mailed ahead by five, I'm repacked by eight, and asleep by nine. Then it's up at 6a.m. and back at the trail head by seven thanks to a ride from the motel owner.

One by one I ascend and descend the passes: Glen, Pinchot, Mather, Muir, and Selden. It's up three thousand feet down four thousand, up four thousand and down three thousand. Each pass is ten to twenty miles apart. Each has views of humongous cirques ringed with twelve thousand to fourteen thousand foot speaks. The Himalayas? Who needs them when one has this to hike through! Bridges on the creeks are nonexistent. One has to boulder hop when you can, wade when you can't and several crossings are quite scary. There's water, water everywhere, just like in the Olympics. And the lakes! Low-level ones ringed with forests, midlevel ones ringed with moss and shrubs, and high-level ones ringed with talus slopes and snow.

Muir Pass is the only mountain place I have ever been where not a tree is in sight as far as you can see in any direction. "Austere" doesn't do it justice. It is quite obvious the area hasn't changed a whit since the ice age waned. An octagonal stone hut with a conical roof is here, built by the Sierra Club in 1932. I decide to spend the night here, legally or not, who really knows or cares. An overzealous ranger has scattered logs, brought up by horses from three thousand feet below, into the talus. I regather them, have a fire in the fireplace, and left this note: "Mountain Dave had a fire in this hut here in honor of its builders who intended it to be that way." Two lakes glow like gold ingots far below. On the descent to Selden Pass to the Vermilion Valley Resort ferry landing, the bugs are so numerous and vigorous they become a thing of legend. Breaks are brief to say the least! It's a race to catch the once a day six-mile ferry, and I make it with five minutes to spare.

VVR is like a wilderness Shangri-la. The first night in a tent cabin, and the first beer are free for PCTers. Butch and Peggy (who really runs the place) are amicable hosts. I pig out meal after meal but it does nothing to fill the empty pit formally known as my stomach.

I've lost twenty pounds, even though I took to eating eighteen-ounce packages of Oreos as a midmorning snack. Damien, Florian, and Let It Be are here, and we gather for the camaraderie of the nightly VVR bonfire. Let It Be has a new girlfriend with the trail name Skunk.

After dropping one hundred dollars in meals alone, it's time to move on. Silver and Donohoe Passes are more subdued than the previous six, barely passing above timberline. The first thunderstorm of the journey greets us at Donohoe with lightning all around, but the downpour holds off until evening when I'm in my tent. The tent leaks like a sieve, but I'm prepared, I put my down sleeping bag into a bivvy sack and place it in a large garbage sack. It works and I stay dry.

By morning the storm breaks. I try to keep up with Let It Be and Skunk, but to no avail. They are headed for the "barn" at Toulumne Meadows and get there several hours before me. Toulumne meadows is measured in square miles instead of acres and is welcomely flat.

The tent store is hiker friendly offering 20 percent discounts to cresters, but the lodge is not. We can only have showers between 1:00 and 3:00 p.m. Desperately wanting bacon and eggs, I am forced to enter the dining room smelling less than pretty. The waitress expects me to share a table with some pink, freshly scrubbed tourists but I protest.

"They don't want to eat with someone who hasn't bathed for a week," I explain to her.

She seems perplexed at what to do when suddenly Damien and Florian yell at me, "Over here, David. You can sit with us. We don't care if you stink."

The waitress continues to insist that I must sit where she says because their table is full, so I demand to see the manager. Sensing a bad scene if I'm not allowed to sit with my friends, the manager is diplomatic and tells the waitress to make an exception. Boy, do I get a Nurse Ratched look, but I quickly calm down once breakfast is served.

And so the Sierras are history. Beyond a doubt it's been a most memorable backpack in a lifetime full of backpacks.

PACIFIC CREST TRAIL PART 3: NORTHERN CALIFORNIA

Northern California starts out with a bang. In my first night's camp out of Tuolumne Meadows, a bear sneaks up behind me as I set up the tent. He gets about twenty feet from my food bags before I see him. My immediate instinct is to attack and defend my food. I scream and start tossing rocks and sticks at him. He seems startled at my actions and backs away slowly. I keep the pressure on and he saunters off. I decide to sleep with my food because of all the stories of bears compromising food hanging strategies. Hanging food is now illegal in most California National parks. It basically is considered bear baiting as it does nothing to deter them from entering a camp to try again and again to obtain hung food until they finally succeed. A new program using rubber bullets to instill the fear of man back into bears is now being implemented.

I hike for a few days with Frank and Linda, sixty-ish PCT section hikers, who are doing about three hundred miles in a month. When I tell them I'm a thru hiker they warn me that in Northern California, many thru hikers quit. The reason? It's kind of a scenic let down after hiking through the Sierras. Much of the next 250 miles is plain old non wilderness national forest. They also tell me that fewer people have completed the PCT in a single season than have climbed Mount Everest. It sounds plausible since only about five hundred people have thru hiked the PCT. I use this knowledge as an incentive to reach a rarely accomplished feat.

There are some surprises in N. California. I hate to say it Washingtonians, but the people of California saved their forests from the bane of clear cuts. When I climbed to a ridge top along the trail, I expected to see the same old familiar patchwork of clear cuts so commonly seen from mountaintops here in the NW. I was astonished to see none! The forests looked as if they had never been logged.

Perplexed, I visited a ranger station and simply asked, "How come?"

The reply? Sierra Club—people got involved and legally forced the logging companies to adopt the practice of selectively cutting mature trees by helicopter. Logging access was restricted to a few miles of double dead end roads along ridge tops where erosion problems would be at a minimum.

Northern Yosemite is flat at first and dotted with lakes. Then a gentle rise past timberline to rounded minor volcanic peaks ensues. Granite is interspersed with basalt for the first time since Mexico, a sign of change to come in the geology along the trail. Lake Tahoe is now prominent to the east. It's midweek, and I have the trail to myself. Florian is moving ahead on her own. She's upped her mileage to twenty-five a day (I'm doing eighteen to twenty-two). Damien has dropped off the trail to nurse a foot injury. Let It Be and Skunk are somewhere behind me. At Echo Lake Resort, the owner is not only unfriendly, but downright rude, refusing to take the time to offer information on a rumored taxi ride to Lake Tahoe. So I road walk seven miles before hitching a ride to S. Lake Tahoe for the first day off in two weeks. I get hold of the taxi service and get a ride back to Echo Lake at a substantial discount for PCT hikers.

Flower season is in full swing. In drier meadows, aptly named mule's ears (a large sunflower) abound. In wetter meadows, it's corn lilies. The regulars are also present Indian Paintbrush, Lupine, Fleabane, Shooting Stars—but no edible glacier lilies! The trail traverses through a fir forest. Douglas fir is now common, along with White fir and a new tree, the Red fir. At lower elevations, Jeffrey pine and Western White pine rule. Partly to pass the time, I have started a game of guess the tree. I attempt to identify trees by looking only at the needles. Then just at the cones. Finally, just the bark. The weather has continued to be perfect (it will end up raining for three hours during the entire hike through North California)

Sierra City, the next supply depot, is a gold mining town founded around 1850 during the California gold rush. I spend the night in a hotel in bad need of repairs. The floor is so unlevel, the doors to the rooms swing open on their own.

Due to the subdued landscape, I'm making great time and am close to being right on my one-hundred-mile-a-week schedule. Other than the scenic volcanic Sierra Buttes, there's not much to see from Sierra City to Lassen Peak National Park. On the descent to Belden, the last supply town before Lassen, poison oak becomes the most common plant along the trail. Avoiding all contact is next to impossible, but I never develop a rash. I decide that I must be among the one in four people immune to its effects and stop worrying about contact with the stuff.

While pigging out in the only restaurant in Belden, Damien pops in. He's putting in thirty miles a day trying to catch Florian who is about five days ahead. There is simply no way I can keep up with him while hiking, so we shake hands and I wish him the best of luck for the rest of his California journey. He is dropping out at the Oregon border to tend the grape harvest at his winery. An endless five-thousand-foot climb ensues out of Belden on a twelve-mile PCT detour due to severe washouts.

Lassen Peak is a surprise. Although I have climbed it before, I had no idea it was so thermally active, far more than all the other volcanoes I have visited. Boiling springs and fume roles make me feel as though I'm in Yellowstone instead of N. California. Even an entire lake is boiling away. When I sign the PCT trail register in Lassen Park, I notice that the two hundred people who started the trail in Mexico has dwindled to less than seventy-five, and Lassen is only the halfway mark. About halfway through Lassen Park, I have my first real injury, a knee blowout. I start out as usual one morning, but within a hundred steps I am stopped in my tracks by an excruciating pain in my right knee. Just a fluke, I instantly hope. The pain returns several steps later. Nope, it's not a fluke.

Panic begins to set in as I realize I'm twenty-four miles from the nearest road. How will I get out if I need to? Even worse, will this be the end of the trip? I try using a walking stick. No dice! More severe pain. I'm shocked by the intensity of the echo from my scream of pain. I begin to feel the first dismay of the journey, the first feelings of doubt about completing the trip. I decide to try wrapping the knee with an elastic sport bandage from my first aid kit. I gingerly try walking on the bandaged knee. No pain! I anxiously hike to the next milestone in the guidebook. Still no pain. I end up putting in about an eighteen-mile day. As a precaution, I leave the bandage on for three straight days. I wince in expectation as I weigh down the unbandaged knee, but the pain is mysteriously gone for good. Trail magic must be at work.

At the next "town," Old Station, the teenage daughters of the general store owner have taken up the trail angel role by providing water stashes along Hat Creek Rim, a twenty-four-mile waterless lava flow escarpment half burned off by a forest fire. Hikers are few and far between here as has been the case for all of N. Cal. I take the time to explore several volcanic features such as Subway Cave, a lava tube very similar to Ape Cave on Mount St. Helen.

At Burney Falls State Park, I lack the three-dollar cash entrance fee, but the ranger smiles and says, "Oh, we'll work it out somehow" and lets me in. There is a campsite reserved for PCTers in a very quiet corner of the crowded car campground. That night, more trail magic is at work. A woman just happens to be giving a slide show on her PCT hike at the park amphitheater. It pumps me up and I walk away utterly determined to finish the Journey.

Next up is the Terrible Section O, infamous for blow downs so severe many hikers abandon the trail for a road walk. Trail maintenance crews have just finished clearing large sections of the trail, but ten-foot high blow downs still abound.

My boots are finally shot as I get to and cross I-5 at Castella State Park. The uppers are separating from the soles. I solicit a ride to a Shasta City boot repair shop from Ted, a retired fellow. He even treats me to breakfast. I offer him cash for his kindness but he absolutely refuses to take it. The boots are unrepairable, so I mend them with shoe glue and pray that they last another one hundred miles to

Seiad Valley where a new pair of Red Wings will be waiting at the post office, sent by Joe and Kathy Weigel, my emergency trail resupply angels. Several letters from PWCers await me at the post office with twenty-dollar bills for spending money. What a lift they are to my spirits. A poem is posted at the free PCT campsite about the trials and tribulations of hiking the PCT. It ends by saying that when you feel like quitting to remember that you're not a failure unless you fail to try. I will end up repeating this line many times over the next two months. Views of Mount Shasta now abound as the trail swings on a 250-mile arc around its western side through Castle Crag State park and the Marble Mountain and Trinity Alps wildernesses. A large forest fire develops in the Marbles and I fret about trail closures as thick smoke begin to blanket the area, but I make it through just in time.

The last resupply station in California is Seiad Valley. As I descend into Seiad Valley along a dirt road, there are large piles of bear poop every hundred yards. The blackberries are ripe, the bears are feasting and the selfish beasts haven't left nary a one for us hikers. One restaurant in Seiad offers an infamous "pancake challenge" to PCTers. Eat five and you get them free do eat one along with a giant Denver omelet. That night I have the best steak dinner of my life at the Wildwood Inn, the hands down best bar along the PCT. It's Saturday night and the place is crammed with locals, a few PCTers and several gregarious gold miners who have plenty of stories swap with me as we sit on the back porch waiting for our dinners. I could add more about what else goes on at this great bar, but I'm sure it wouldn't pass the censors. Later, in my tent, I'm awakened several times by the sound of bears thrashing through nearby blackberries.

Two days later after another grueling 3,500-foot climb in brand new Red Wings (nope, no blisters!), I'm standing at the Oregon border trail register elated after four months on the trail. I notice that Damien has caught up with Florian. All is right with the world! There's 1,650 miles down and a thousand to go!

PACIFIC CREST TRAIL PART 4: OREGON

A giant stormsweeps through the area the day I enter Oregon (September 1). It drops into the thirties and the rain mixes with snow. The same day, I suffer my second major injury, a twisted back, not from hiking but simply bending over to pick something up. It's not so bad the first day, but the next morning, perhaps due to the damp, cool weather, I can't even get to my knees without severe spasms of pain in my lower back. It takes over two hours to break camp, but the pain subsides considerably after I hoist my pack and start out.

Ashland is the next stop and I decide to take a full day off, hoping the pain in my back will go away with rest. Most PCTers seem to like Ashland, but I find myself feeling very negative about the place. In retrospect, I regret this negative mood, but I was full of anxiety about my painful back. My impression at the time was that it's an island of Southern Californian culture full of trendy boutiques, expensive ethnic restaurants and various and sundry new age causes.

Apparently, Ashland was chosen because it has the Shakespearian Festival, the only thing resembling "culture" for hundreds of miles around. Ashland is also a college town with the attendant bicycle shops coffee shops and organic grocery stores. Toss into the mix the usual garish strip malls along the Interstate and Ashland is pretty much summed up. Paying seventy dollars for a room without a bath doesn't help any. The next night I transfer to a hostel. Still no bath in the room, but it's only fourteen dollars. My negative mood expands to sap my will power to continue on. I seriously consider quitting the trail, but then I remember: "You're not a failure unless you fail to try and press on."

The next fifty miles are mostly cow pastures full of you know what. When I encounter one herd I yell at them, "McDonald's! Burger King! Veal! Run for you lives! Don't you know they're going to eat you! They just look at me with their big brown eyes, batt their long eyelashes, chew their cud and seem to have a "What's with him?" look on their face. Calves frolic in the meadows.

I'm now leap frogging the trail with another David (trail name: Caboose). We are apparently the last thru hikers of the season. No one has passed us for over two weeks. Mount McLaughlin is the first Oregon volcano on the trail. There is still plenty of snow on its north side from last winter's record snow. I've been told by many people that the North Cascades aren't going to melt off this year and that nobody is going to get through. I've decided not to worry about it until I get there.

When I get to Crater Lake I sneak past the entrance station to avoid paying the $10 entrance fee. I should have known better. Another hiker later tells me he did the same thing but was caught and told, "Hey, you get in free if you're a PCT hiker so there's no need to sneak in"

I whine at the cashier about the super high prices for food at the store. She looks downcast as she says, "Hey, don't blame me I don't set the prices." I feel terrible and apologize profusely. Her spirit brightens remarkably and she helps me get a ride to the post office to send off a parcel. After dinner at the lodge I go over and gaze over sacred Crater Lake. I am mesmerized by it and just can't leave. I stay until dusk and illegally pitch my tent on the rim when I'm sure no one can see me. I pay for it later that night. I had heard that weird things happen here and sure enough they do.

I have the most vivid dream of my life. I was an adolescent Indian boy and the tribe's medicine man was putting me to bed inside a small cave. He sprinkled a special herb on my chest and placed a special blanket over me made of tree bark. Then he admonished me not to take the blanket off no matter what happened that night. I drifted off to sleep only to be awakened by a bright flash that turned night into day. Several seconds past. Then an enormous blast of hot air came into the cave. Hot cinders rained down on me. An exposed arm was burned and I pulled it under the blanket as I screamed in terror. The blast seemed to last for hours. When it finally subsided I arose to find the forests completely gone and a black goo covering everything. Most of the tribe had perished including my parents. Other survivors wandered around in a muted daze. Then I awoke, but I was too scared to open my eyes for a long, long time. Guess that will teach me not to camp on the rim of Crater Lake!

Resupply strategy becomes critical for most of Oregon. The PCT passes through only tiny resorts with very limited supplies such as beer, potato chips, and candy bars. After a small side trip to Diamond Lake resort for a good breakfast, I pass by Mount Thielsen, a nearly vertical volcanic plug spire, nicknamed the lightning rod of the Cascades. The weather has returned to perfection with dozens of clear sunny days in a row. Oregon is nearly flat, trail wise, compared to California. Twenty-mile days become easy, and I'm cruising right along. There's not much to say about Shelter Cove resort except that they have showers. The Three Sisters wilderness begins near here and the trail passes by numerous nearly bug free lakes. Due to thick forests, views of the Three Sisters, three medium-size volcanoes in close proximity, are rare until near Elk Lake Resort when both Broken Top and Bachelor Butte also appear. Volcanoes are bustin' out all over!

I spend a night in a one room cabin for ten dollars at Elk Lake, run by a bona fide chef, and have a custom made breakfast. I pick up a package sent from Ashland, and it's onward again through a phalanx of volcanoes and surrealistic bizarre volcanic features. The most bizarre of all has to be Belknap Crater. Here, lava oozed out like toothpaste and spread out for many dozens of square miles instead of building up vertically. The flows happened only a few thousand years ago, and trees have yet to successfully pioneer in them. The result is an erose trip through mile after mile of black frothy lava. One gets the impression of a giant black amoeba that has devoured its way through a thick green forest. It's a great foot relief when the trail finally returns to its cushion of soft spongy soil.

Mount Washington and Three Fingered Jack are volcanic plugs similar to Mount Thielsen. I had originally planned to climb them but I've rationalized my way out of it by saying I'm prioritizing. Getting to Canada before the snow piles up is just more important right now. I also tell myself I'll be back someday while fully realizing that someday may never come. I'm torn, and the PCT wins.

Mount Jefferson is stunning. I recall vistas of it so impossibly beautiful that all you could do is stare and shake your head. The first massive bona fide glaciers of the trip are on its slopes. It's mid-September and I have the place to myself as I pass through Jefferson Park. Glacier fed creek crossings become a problem for the first time since the Sierras. Mountain huckleberry and blueberries are now abundant and I browse as I walk along. Lakes become abundant once again on the northern slopes of Jefferson.

At Ollalie Lake I meet a hiker that I had dinner with way back at Kennedy Meadows. For a variety of reasons, he is now hiking southbound through Oregon. We split the cost of a cabin and I resupply from a package mailed from Crater Lake. The owner is also opening a package sent by PCT hikers who have called to say they are not coming. It's first come first serve as I rummage through their package to supplement my own supplies. Have I mentioned hiker boxes? These are supplies left behind by hikers for reasons of their own mostly related to weight. The resorts and post offices along the PCT maintain them along with official trail registers. After a while, you get to recognizing names in the registers, and start looking for them in each one. There are around fifty names in this year's register.

As I approach Mount Hood the weather begins to deteriorate and by the time I reach Timberline Lodge a raging ground blizzard along with heavy snow squalls develops. I decide to get a room but the place is booked. I find a flat sheltered area in a clump of trees less than 100 yards away and make camp. I can see the lodge from my tent and I try not to feel bitter knowing the guests there are warm and comfy in their cozy rooms while I'm out here shivering away, but to no avail. The next morning I go in for breakfast and laundry, determined to hate the place, but yowie wowie, being a builder, I immediately fall in love with it. The design and craftsmanship are exquisite. The place was built by unemployed craftsmen as part of the Civilian Conservation Corps during the Depression. It looks like the builders and artisans were told, "We don't have a schedule or a budget. Just do the best job possible with the finest material available." The place is a study of perfection! It seemed like the cheerful, helpful staff knew they were very lucky to work here. They had the attitude that if we don't know something, we'll find out. Breakfast was expensive, but well worth the price with a complementary vial of wild blackberry nectar.

The weather improves slowly over the next two days. Now I am presented with a choice: be a purist and stay on the PCT or take the alternate Eagle Creek trail. Even the guidebook says not to miss Eagle Creek with its myriad waterfalls. The PCT goes through viewless forest. It's not really a very hard decision to make. Eagle creek it is.

The most impressive falls is Tunnel Falls, perhaps two hundred feet high. It is approached on both sides on narrow ledges with chain handrails. A tunnel has been carved out behind the falls and it is a unique feeling to see and feel massive amounts of water passing directly over your head. Ramona and Punchbowl falls are also very picturesque: Ramona because the water flows over dozens of ledges about one hundred feet wide and Punchbowl because the water drops into a very large pool surrounded by impossibly lush greenery. There is certainly no regret for taking this detour from the PCT.

I giggle with glee as I finally spot The Columbia River and plod on into Cascade Locks. Washington here I come!

PACIFIC CREST TRAIL PART 5: WASHINGTON

Glacier Peak

As I cross the Bridge of the Gods over the Columbia River into Washington, the bridge keeper flashes me a big smile as she wishes me good luck. The weather holds for the first week in Washington as I pass through Indian Heaven and Mount Adams wildernesses. Mount Adams is a bit of a disappointment because a forest fire's smoke obscures most views of it.

The night before entering Goat Rocks Wilderness it starts raining. The next day as I pass the six-thousand-foot level it begins to snow and it is tense, anxious hike through the fog. The PCT traverse's the highest ridge in Goat Rocks and I am worried about losing the trail as it crosses permanent snowfields and getting lost. At each and every trail junction, I seriously consider bailing out to logging roads far below. The really bad weather holds off until I near White Pass. It begins to rain heavily and I get fire hose soaked before getting to a welcome warm and dry motel at the pass. I watch a Disney movie named Iron Will on TV. It's about an eighteen-year-old who wins a long dog sled race against all odds. It's just what I need to improve my spirits while looking out the window at the pouring rain. The next day, the rain continues unabated and I wait until checkout time before pressing on. Surely it can't rain for three days straight I rationalize. It rains even harder, and I'm soaked again within three hours. I pitch my tent around 3:00 p.m. to wait it out. By morning the storm is past and I get to enjoy lake filled William O. Douglas wilderness.

Next up is the Shame of the PCT—a forty-mile stretch of clear cuts that disgusts every hiker who passes through it. As I approach Snoqualmie Pass, I get drenched once again as I push through thick brush that has overgrown the trail. The woman trail angel who runs the Time Wise Grocery takes pity on me and makes several calls to find lodging for me. She books a room at a nearby Best Western at a 40 percent discount and even takes time off to drive me there. The weather forecast is for a spell of great fall weather. I pass through Alpine Lakes Wilderness as the fall colors begin to intensify due to cold frosty nights. I had never hiked here before (except for the Enchantments) and am surprised at the astounding beauty of the place. The trail passes between high jagged peaks and numerous large lakes far below me. Recent snows cling to the north sides of the mountains. I have the trail entirely to myself.

At Stevens Pass, I am greeted by PWC members Alyce, Dan, Gary, and Alice who have brought me a resupply package. At times, I feel like I'm dreaming as I eat lunch with them. Their presence is a delight, but it doesn't seem quite real after a five-plus month journey among strangers. They are shocked by my emaciated appearance.

I'm apprehensive as I pass through Henry M. Jackson and Glacier Peak Wildernesses and the trail goes through passes around seven thousand. This spell of good weather can't last forever in a Washington October and a large storm would be catastrophic here. The fall colors approach an unearthly nature—bushy cushions of crimsons golds and violets blanket every hillside and have my senses reeling with many views of heavily glaciated Glacier Peek to top it off. The descent to the Stehekin road goes smoothly. The bus to Stehekin is no longer running, so I start hoofing it down the ten-mile road. Several hours pass before the first car goes by and I hitch a ride. I ask the waitress at the restaurant if any PCTers have been through lately.

"Yeah," she says. "A gal by the name of Florian and a guy named Robert. They had breakfast here and headed out together about a week ago."

So they finally did meet! I check the register at the post office. Caboose hasn't been here yet and I wonder if he made it through that storm back at Goat Rocks. I do the usual resupply stuff and it takes about 3 minutes to hitch a ride the next morning back to the trail head.

Strange as it may seem, losing a watch may have saved my life, or at least spared me much discomfort. I lost it one night out from Stehekin. I must have yanked it off my arm while putting on my pack in the predawn darkness. The next night, I had no idea what time it was when I awoke the next for the push into Harts Pass. It must have been around 2:00 a.m. since I hiked around seven miles before dawn. By then my feet were so cold that I started a fire right on the trail to warm them up. I noticed the sky to the south was a steely gray.

It was an ominous sign of things to come. As the day wore on, the wind blew harder and harder, the sky grew darker and darker, and the clouds moved lower and lower, obscuring the mountains to the south. Since I was well above timberline, I pleaded with the storm to hold off, for this was definitely not the place to be in one. I was quite relived when I got to Harts Pass and its access road. I was there no longer than five minutes when the snow came in great gusts, dropping visibility to an eighth of a mile in mere minutes! If I had not lost my watch, I'd have been four hours back on the trail with no shelter to speak of in a raging blizzard.

A car came by and I flagged it down to get a weather report.

"A quickie," he assured me. "It's supposed to last only a few hours with little accumulation."

Famous last words, for sure. I was going to ask for a ride to Mazama if he said it was to be a bad storm. I watched him drive off with a sinking feeling in my gut. He was to be the last car to drive to Harts Pass for the next seven months! I left a note on the Ranger station door as to my plans and started out for the last leg into Canada now a mere twenty-five miles away, but thought better of it and camped near the ranger station instead. It snowed eight inches that night, but I could still make out the trail the next morning and decided to go on.

"You're not a failure unless you fail to try!" I kept repeating. I knew there was a yurt about five miles away that I could use as an emergency shelter. Snow squalls came and went and I decided to have lunch and warm up at the yurt.

After eating, I hiked another two miles to a ridge top when a little voice inside me said, "Man, are you stupid or what! Here you have a warm well stocked yurt to spend the night in and you're throwing it all away to spend a miserable night out in the snow. Turn back and spend the night there, Dumbo! There's still plenty of time to make it to Canada."

So turn around I did, and it was the right decision. The minute I got back to the yurt, phase two of the blizzard came in. It snowed hard for the next eighteen hours. The yurt shuddered in ninety-mile-an-hour winds all night long. I admit it, I was terrified that the yurt was going to collapse. "It's going down, it's going down!" I kept thinking with every giant blast of wind. I even placed my jacket next to me on the bed so I could keep warm in the collapsed rubble, if it came to that.

The next morning, I still wanted to go on. I could see a section of the PCT on a hill from the Yurt's window. I hoped against hope that the wind had scoured the slopes bare. "You're not a failure unless you fail to try!"

Less than a hundred yards from the yurt, reality finally set in as I waded through two to three foot drifts. There was simply no way I could plod through the snow on the final steep up and down twenty miles to Canada. It would be hard enough to hike the relatively flat five miles back to Harts Pass. Very occasional indentations in the snow marked the trail back, and heavy snow squalls kept me guessing as to whether the blizzard was coming back. It was an exhausting stressful, depressing hike back. I decide to break into the Ranger station and await a rescue ride from some intrepid four-wheel driver, but it was well secured and I couldn't manage to do it. I started post holing the twenty-mile road to Mazama and passed a large rock slide blocking the road. About dusk I finally passed the snow line and made my final camp. I arose well before dawn and began hiking the final twelve miles to Mazama. Around dawn, a car with four retired gents in their eighties picked me up. "Anyone who's hiked as far as you did deserves a free breakfast," they said after I told them of my PCT journey. A kind hearted young woman gave me a ride to Bellingham where I caught a bus to Seattle. From there I took the ferry to Bremerton, and the foot ferry to Port Orchard. Alyce picked me up, I took a shower at her house and she drove me to the PWC Halloween party. The end.

It took a while, but I finally realized that being stopped only twenty miles from my destination was the best of all possible outcomes. I wanted in the very beginning for this journey to be a lesson and an experience. If I had completed the PCT, I feel that I would have lost sight of this and viewed the journey instead as an accomplishment. I learned that perseverance is a virtue, but that as that philosopher Clint Eastwood once said, "A man's got to know his limitations." If you arrogantly challenge Mother Nature and spit into the wind, the only one who suffers is you.

I would like to thank all the trail angels and PWC members who helped me on this incredible six-month journey. I will forever be humbled by their generosity. The journey would most certainly would have lacked much meaning without their physical and mental support. And if you ever get the opportunity, don't you dare pass up the chance to hike this wonderful, wonderful trail of trails, the Pacific Crest Trail!

MIKE URICH EUOLGY

The following eulogy is carved into a large entrance sign for a Pacific Crest Trail hut located in northern Washington:

> The mountain gods from seats on high
> Rejoiced to see Mike Urich die
> And at his death gave this decree

"To all who pass here know that we
Entrust to Big Mike Urich's hands
These camps, these trails, these forest lands
To rule, protect, to love and scan
Well as he did while mortal man
And deal out sentence stern and just
On those that violate his trust"
Stranger beware, leave not a fire
Foul not Mike's camp, rouse not his ire!

THE RIGHT PLACE AT THE RIGHT TIME

"I'm sorry, but the campground is full on those dates."

I had just tried to add Linda to my camping permit in Havasu Canyon. She had called a few days earlier wanting to go on my one-hundred-mile Grand Canyon trek. I had tried to inform her that it was for experienced canyoneers only and that it was a very dangerous hike in a remote corner of the Canyon with many miles between water sources. If anything happened at all it would be days before help could arrive. "Cool, sounds like fun," was her response, even though she admitted that she had never hiked in desert conditions before.

So now I had the perfect excuse to not let her go. But somehow, I just couldn't. I admired her pluck so I decided not to say anything and just deal with the Havasupai Indians when they demanded to see our permit. It said on mine we would have to hike back out if we had no reservation.

The day of departure finally came and off to the Canyon we went. We started out with 6 people, with four of us dropping out after the first camp. Only Linda and I would continue on from the South Rim of the canyon to Supai, We had all kinds of dire warnings on our permit telling us to beware of this and that, but one comment from the ranger was revealing. At the end of our itinerary he had written, "Yes, yes, YES!"

We hiked from Hermits Rest to Boucher Creek and then on through a half dozen side canyons known as "The Gems"—Ruby, Turquoise, Sapphire, Jade, Serpentine, Quartz. Each of these canyons went back into humongous amphitheaters and as Linda commented, it should be called the Grand CANYONS. Any one of them could have been a National Park in their own right, all could have swallowed up Zion whole. Ten miles between water sources was the norm. We took siestas between eleven and three to conserve water. Although there was a severe drought, the seasonal water sources were "running." Actually they were the barest of trickles.

On our way down Bass Canyon toward the Colorado, we both ran out of water. We came upon a pothole with about ten gallons of water in it. I never thought I'd see the day that someone would drink water that I would not. This water was stagnant and full of green slime with numerous pollywogs swimming about. Linda filled her water bottle and it looked like lime Kool-Aid. She took a big swallow and offered me some. I declined, deciding instead to hike five waterless miles to the river. The pollywogs were grateful to me, I'm sure. Linda is still alive and well and I am humbled.

The next leg of our hike took us back up to the rim of the Canyon and then to Topocoba Hilltop for a descent to Topocoba Spring and on to Supai on the Havasu Indian reservation. We each took five quarts of water with us, hoping to resupply on the rim from natural "tanks"—depressions that filled with melting snow in the spring, or from campers at a campground. When we reached the rim we were down to less than two quarts each. The extremely remote campground was deserted. By morning we were down to less than a quart each. We hiked to the nearest "tank" and it was bone dry. So was the next one. Linda was getting worried. Having been in tight jams on numerous occasions, I felt an odd sense of serenity. I just KNEW that Providence would provide, as it always had before.

"Not to worry," I said. "We'll hike to this road junction on the map and somebody will eventually come by." If they didn't we were faced with two options, both waterless: a thirty-five-mile hike back to Hermit's Rest, or a twenty-four-mile hike to Topocoba Spring, not knowing if it was flowing. As we were hiking toward the junction, Providence provided in the form of two reckless dudes who came flying up from behind us and had obviously never heard about Mothers against Drunk Drivers. After handing us each a beer (it was

10:00 a.m.), they offered us a ride to the junction, only a mile away, or back to Hermit's Rest. They had about ten gallons of water on the back seat and offered us as much as we could carry. I told Linda she could return to Hermit's Rest, knowing I had no camping reservation for her at Supai, but that I wasn't ready to call it a trip yet and was continuing on. She wanted to continue on too. The two drinkers offered us another beer, but we had to swear first that we wouldn't toss the empties. I noticed the Nature Conservancy cap the driver was sporting, and said, "sure, no problem." Then on we went after Linda "flashed" them as a reward. What can I say? She was that type of girl. She was also becoming irritated with my pace, accusing me of deliberately going slow to frustrate her. She wanted to hike as fast as possible to Topocoba Spring and just get the twenty-four damned miles over with as quickly as possible. I said I was already hiking as fast as my short little legs could. She said, "Well, fine then, see you in camp," and off she went.

I hiked the twenty miles to Topocoba Hilltop along the rugged four-wheel drive road and nary a car went by. Waiting for someone to come by with water would have been a bad idea, it turned out. Thank God for the two drunks! By this time I had lost Linda's track and had an uneasy feeling that she was off track. We were now in the middle of one of the most remote areas in the entire lower forty-eight states. Getting lost was NOT an option.

I had no choice but to hike on alone to Topocoba Spring. As I neared the spring I could see that someone was camped there, with a campfire going. I knew it wasn't Linda because her track was not on the trail. To my dismay, I could see horses and realized the Indians were camped there. They knew I was coming because I had obtained special permission to hike the seldom used trail. How they would react to no permission for Linda was weighing on my mind, so I tried to bypass the spring without them seeing me, hoping they would leave by morning. Suddenly I heard someone yell out, "Over here, over here! They had spotted me. One of them left camp and headed toward me. "We found your friend! She's camped on the rim above. We were catching wild horses. We knew you were coming and were watching for you. She was headed the wrong way so we raced up to her to tell her. She said she was too tired to hike any further and would come down to our camp in the morning. You can camp with us tonight. Can I take your pack for you?" He was very impressed by the distance we had hiked from the South Rim. "No one has hiked to Topocoba Spring from the Grand Canyon in at least five years. We are eager to hear your tale. My name is Raphael, what's yours?"

When we arrived at the spring he introduced me to his friend Cliff and his teenage son. He showed me the two horses they had captured and said they hoped to catch more the next day. A stew was cooking away on the campfire. After dinner, the mandatory peace pipe was passed around. They said they were the rangers for the reservation, but did not ask for my permit, saying instead to check in when we got to Supai.

Just as twilight was approaching, Raphael spotted two wild horses headed down the trail toward the only water source for who knows how many hundreds of square miles around. "Shh! Shh! Hide behind these rocks," he whispered. The Indians crept into well-rehearsed places and to my utter astonishment, I began to witness a rarely seen event unfold. I watched from my vantage point as the two horses inched ever closer, the Indians and myself hidden from their view. Once they passed a certain point, Raphael's son ran from his hiding place. Now behind the horses, he yelled as loud as he could to scare them forward toward the spring while simultaneously pulling a makeshift barbed wire gate across the trail. The horses were now trapped by the great cliffs surrounding the football field sized amphitheater containing Topocoba Spring. A large talus field led into the amphitheater, the only feasible escape route being the now blocked trail.

Suddenly all hell broke loose and stayed loose for the next thirty minutes. The two horses began to run in wide circles inside the flat amphitheater, looking for a relatively easy way out. The three Indians took up positions at the weak points. The horses would race up to them one at a time at full gallop, rear up, and whinny as loud as they could. The Indians waved their arms and shouted "HEAH!"

as loud as they could, and each time, the horses backed down. The Indians all glanced MY way and it was obvious they expected ME to do my part. And sure enough, the larger of the horses galloped straight toward me, while I was thinking, “I didn’t sign up for THIS! I’m about to be trampled to death. Never thought I’d die THIS way.”

So I just copied what they did and waved my arms while shouting HEAH! The horse reared up all of fifty feet (well, it SEEMED that high), looked me straight in the eye, gave a nasty snort and much to my amazement, backed down and ran back into the amphitheater. He repeated the process two more times. Meanwhile the other horse had said, I’ve had enough of this crap and ran right over Raphael’s son, on down the trail and jumped over the barbed wire fence after stumbling and falling several times. The bigger horse continued to run in circles. Raphael, an expert ropes-man, tossed a lariat into the air and it went right around the horse’s neck. But the battle was FAR from over because there was no way in Hades that Raphael could control that horse as it dragged him like a rag doll over rocks as big as refrigerators with him hanging onto the rope for dear life.

This continued for what seemed like an eternity until Raphael was able to coil the rope around a large rock. The horse now ran until his rope ran out, twisting him to the ground. Raphael now worked furiously, trying to make the loose rope ever shorter by coiling it around large rocks. This was the critical time for preventing the horse from injuring itself. But despite their efforts, the horse nearly died when he entangled himself in the rope, tightening it around his neck so that he could not breathe. After fumbling unsuccessfully to uncoil the rope, Cliff quickly took out a knife and cut the rope, and one could hear a great gush of air go down the horse’s throat. Rapheal quickly lassoed one leg and then another and before you knew it the horse was hog tied and immobilized on the ground. Raphael crept closer, wary of the horse as it tried to kick free. He slipped a halter over the horse’s nose and neck, and attached a short rope from it to a nearby tree. He then cut the ropes entangling the horse’s feet, but the horse would not get up.

"He's in shock and is going to die if we don't get him to stand up!" Raphael desperately shouted.

He took a whip out and gently flailed it across the horse's belly. The horse twitched, and flinched, but still did not get up. He repeated the process several times to no avail.

"I hate to do this," said Raphael, "but I MUST."

He went over and kicked the horse in the head, and THAT did it. The horse immediately arose.

And then all was calm again. The horse knew he was beat, and was not afraid because he could see other horses around that were not the least bit concerned to be among humans.

What can I say? I was now bonded with the Indians because I had helped them. They told me I could stay on their land for free from now on, that they would pass my name on to all fee takers so I could avoid paying the customary camping and entrance fees. I pushed it.

"What about Linda? I don't have a permit for her," I blurted out.

They said "Yeah, sure, her too!"

So much for that long-anticipated problem!

That night around the campfire they told me several of the tribe's legends, including how their mortal enemies, the Hopi, had betrayed them back in the 1500s when they showed the Conquistadors how to find their village, Supai. Many people died in the ensuing raid. Another legend I am not at liberty to divulge. They told me where to find the sacred ancient petroglyphs that no white people were allowed to see. Cliff recounted how he had just enough time to climb a tree during the great Havasu flash flood a few years back, and desperately clutched its branches for twelve hours before the waters subsided. Meanwhile, the peace pipe went round and round, as the captured horses stared at us sullenly, their fate as future pack horses sealed.

The next morning, Linda hiked down and joined us for breakfast. We then hiked fifteen miles on down to Supai, visiting the sacred petroglyphs along the way.

After two days of exploring, including a hike down magical Havasu Creek to the Colorado, Linda had to leave to catch an early flight. She later told me she had no problems hitching a ride to Kingman. I was not so lucky. I got a ride to Peach Springs, but still had about eighty miles to go to catch a bus in Kingman. After about three hours, I crossed the street and began accosting drivers at a gas station as they filled up, offering twenty dollars for gas money. Still no luck. Then an Indian came in and I recounted my wild horse story to him. "Oh yeah!" he said. "I heard all about it! Raphael is my cousin. I'll get you a ride." A few minutes later, two Indian ladies showed up who were, shall we say 'supersized', driving a run-down compact car. They had a teenage boy with them. We'll get you to Kingman they said as they took my twenty dollars, but first we have to make a stop." They drove straight to a drive in liquor store and bought a couple of six packs to go. And then on down the highway we roared at one hundred miles per hour. It was a very straight road. They offered me a beer and said, "don't worry we won't scalp you, we're friendly Indians." They took great pleasure in tossing the empties out the window, as one recounted how she use to have a good paying job, but alcohol had ruined her life. Less than forty minutes after leaving Peach Springs, they dropped me off at a cheap motel right next to the bus station and the rest of the journey was pretty mundane.

FEAR AND LOATHING ON THE ARIZONA TRAIL

Sonoran Desert AZT

I have written numerous article for the Peninsula Wilderness Club's newsletter, and have always tried to keep a positive—or at least a neutral tone—in my trip reports.

I figure I'm due for a little negativity.

Now I could go on and on about what a great hike the Arizona trail was, spewing out adjectives describing the most abundant flower display in recorded or legendary history. Or I could wax on about the "awesome" and "far out" feeling of "bliss" and "contentment" that is reserved only for those who have been on a trail for over a month Or I could effuse about unexpectedly seeing both snow and Aspens less than five miles from the Mexican border.

I could describe the eventually cordial meeting between me, PJ—my hiking companion for the first four days—and about thirty illegal aliens concerned, at first, that we might be with the Border Patrol.

But I won't.

Because it is only now, after six months away from that God awful excuse of a trail that I can finally calm myself down enough to write something coherent concerning my angst.

First of all, I certainly don't want to diss the scenic beauty of this six-week three-hundred-mile hike. It pretty much had it all, with the interplay of desert and desert mountain range happening numerous times along the way. I am also grateful for the "historic" wildflower display caused by a "historic" Arizona rainfall several weeks earlier, also ensuring the seasonal water sources would be there when needed.

But compared to other long distance trails, such as the PCT, the seven-hundred-mile-long Arizona Trail is, shall we say, a bit lacking when it comes to trail maintenance and marking. It took me a long time to figure out why.

Let me elaborate!

Now mind you I had been warned that the AZT was a skosh on the unfinished side and was nothing at all like the Pacific Crest trail, which I had previously hiked. Besides, I was quite experienced in off trail travel thanks to my copious mountaineering outings, so my ego scoffed at the idea of this trail being all that difficult.

The second day out we came to a junction with no AZT marking on it, guessed wrong, needlessly dropped 2000 feet, and wasted an entire day just getting back to the correct trail, thanks to a chance meeting with local hikers who knew the way through the myriad of trails in the area.

So now imagine yourself following the trail also used by cows to avoid various and sundry sticker bushes and cacti straight from the depths of hell, with names like "Horse Killer Barrel Cactus"—be sure to carry your blood type in that Camelback instead of water if you plan to wade through the stuff in shorts! You'll need it for the transfusion.

So along comes a fork in the cow path, and guess what! There's no inkling whatsoever as to which cow path to choose! Do we do like Robert Frost and take the cow path less travelled or what?

Imagine this happening "numerous" times a day. It's a fifty-fifty chance each time as you play flip the coin to decide which way to go. And half the time you guess wrong!

After several days on the trail, PJ bailed out for two reasons: She had gotten bad blisters—the usual bane for novice long distance hikers, and she was tired of not knowing where you really are day after day. Luckily, she had a friend who lived less than an hour's drive away from a fishing resort near the AZT. She called her, and just like that she was gone.

Two weeks of guessing which cow path to follow was beginning to get me a bit irritable. The mountain range trail systems were ancient, built in the fifties, with little or no inkling that they were part of something called the Arizona Trail. The connecting lands were basically braided cow paths, especially along a seventy-five-mile stretch called Cienga Creek. I'm sure all the smugglers knew the way like the backs of their hands, but I basically followed the creek via the cow path of least resistance. After a day's rest in Benson and the help of a couple of trail angels, things really went to hell in a head basket. Oh sure, the trails in Saguaro NP, were well maintained but it was difficult without a good map to know which way to go because the AZT wasn't listed on any of the trail markers—just local trails.

After getting off trail for the 531st time in Sabrino Canyon in the Santa Cantillina Mountains, I came upon what the "guide book" says is the best camp on the AZT—Bubbs Pool on a roaring river deep within a desert mountain range. Ah serenity at last! But it turned out that ignorance was bliss. The next day the plan was to ascend some three thousand feet to Romero Pass. From there the trail went down the Canada Del Oro River for many miles eventually connecting to roads going to Oracle, the next supply town along the AZT. To my alarm, the trail quickly petered out and I was then forced to bushwhack up every one of those three thousand feet, the dreaded sticker bushes clinging to and grabbing everything they touched. By the time I reached Romero Pass I was, shall we say, a bit peeved with those responsible for maintaining the trails in the Santa Cantalinas. Occasionally the trail tread did appear before disappearing beneath thirty-year-old desert scrub oak, junipers, and the scratchies. I waded through brush ten feet high at the pass, eventually stumbling upon an ancient metal trail junction sign. I started down toward the Canada del Oro on what became a pretty good trail, but I should have known better by now. After several hundred yards, the trail faded into the Barbs of Beezlebub once again. I eventually reached the river and discovered that the "trail" had been completely washed out and the rocky, boulder—and wet—river bottom was now the "trail" instead. Every once in a while an ancient trail marker would appear on a Ponderosa pine to taunt and remind me that there USED to be a good trail through here.

After many hours of this, many crossings, and one night's camp, I came upon another completely overgrown trail junction sign. The Arizona Trail supposedly goes up another three thousand and then follows a ridge that eventually leads to Oracle. However, the map I have shown that if I continue on down river I will eventually reach a four-wheel drive road that also goes to Oracle. The trail to the ridge looks just the same as the trail going down river with its canopy of sticker bushes. I am not about to fight the stickers up another three thousand feet just to go back down again to the same place that road goes, and I assume the road won't be as overgrown as the trail. I start down the river and eventually come to the road.

So here I am walking down this road a disappointed, seething, scratched up mass of negative hostility. By now the ONLY thing that keeps me going is the hope that there is a ranger station in Oracle where I can vent my rage at their lack of trail maintenance. I begin to fantasize about what I will say as I burst through the ranger station door—my apologies to Clint Eastwood:

> You know, being experienced, I don't mind so much traipsing aimlessly through your sorry excuse of a trail system that in reality is nothing more than a sticker bush hell. But this here Monkey on my back is a bit upset about being dragged through all of that, especially after I assured him that there would be a trail like the map says. So if you will just apologize to old Kelty here, like I know you will, perhaps I can convince him not to jump off of my back and SCRATCH you all to death!

I continue walking down the severely undulating road that does nothing to improve my sour mood. The extreme four-wheel drive road, posted with "closed" signs, crosses the river many times. By now I don't bother with crossing procedures, I just slosh across. Suddenly a caravan is conjured up like a mirage in the desert. Who COULD it be coming down this godforsaken—and closed—road. As they approach, I notice a Forest Service insignia on the lead vehicle. Of all the people it could possibly be, it is the head ranger for the entire Santa Cantalina forest along with an entourage of SUVs. Apparently they had all had a case of office fever and were out four wheeling around just for the hell of it, even going up the Canada del Oro for about a hundred yards to avoid a REALLY bad washout. It was like my ire had conjured them up out of the desert sands, and here they suddenly were to answer all of my trail maintenance queries.

I flag them down. I approach. A window rolls down. The moment I have been waiting for has arrived. I haul back and punch him in the nose! Okay, not really.

"You know," I say to him, "I have probably hiked in at least fifty different National Forests and a bunch of Canadian ones too, and I have got to say that your trail system is the worst one of them all. Why? Because, Buster Brown, you don't HAVE a trail system. It is completely gone. The trees growing where the trail used to be are at least thirty years old, so whomever has been paid for doing trail maintenance is guilty of fraud because I am certain not a dime has been spent on real trail maintenance since they were built back in the fifties. So what have you got to say for yourself for allowing this sorry state of affairs to develop?" I graciously left the monkey's feelings on the matter out of the discussion.

The head ranger babbled on about fires and budget stuff for a while with me interjecting conflicting logic. I proudly showed off my scratches and shredded clothing to emphasize a point. The monkey just growled.

I noticed they all seemed to be smoking and suddenly it dawned on me that hiking trails were less of a priority here than, say, four-wheeling around in a Jeep!

The head ranger finally broke down under my relentless criticism and admitted that it was all his trail maintenance boss's fault—I was astonished that such a position existed—and he agreed with me that the guy should be shot on sight for keeping him totally in the dark about the reality of the situation. No, not really! He gave me the guy's number and told me to take it up with him.

Then he said, "Thank you for your input, and have a nice day."

Then the window rolled up, and they merrily conjured back to where they conjured from.

I had a smile on my face the rest of the way into Oracle even though for some odd reason they didn't offer me a ride. Two kindly senoritas eventually offered me a ride, and dropped me off at the town's motel.

Even today I question whether or not this occurrence took place. I mean, like, what were the odds of them showing up at my peak moment of angst? Perhaps I was vexed beyond sanity and hallucinated the whole thing. Maybe it was sticker scratch fever or a bona fide desert mirage. Alien abduction gone bad because of my bad attitude? Too bad I don't have pictures for evidence.

After a rest day in Oracle and five meals at the only cantina in town, the real deal Sonoran Desert began. Faint, seldom-used ranch roads extended in straight lines for many dozens of miles into the heart of nowhere, the only water sources being widely spaced—fifteen to thirty-five miles—open to the cows windmill pools. Ever notice how they slobber when they drink out of one? Neither did I until this hike. I notice now!

And of course the uncertainty continued heightened by the fact that if you chose the wrong ranch road you would, basically, die of thirst.

So after several days of the ultimate desert communion with the Slobbering Cows, Saquaros and wind mills, I arrive at yet another mountain range with a much lower incidence of getting off the trail— mainly because there was no trail to get off, just an educated guess as to which wash headed generally north. I finally get to the Gila River after several final uncertain hours along a ubiquitous looking ranch road that seems to be headed in the right direction. Maybe. If I'm lucky.

I cross the river on what appears to be an abandoned railroad trestle. Fifteen minutes after crossing it, a train disproves that theory. Now I thread my way up a trailless wash, zigzagging through the cholla, saquaros, and stickerarium galorium, eventually reaching yet another four-wheel-drive road.

A mere three false starts later—defined as at least one mile off trail, one way, and an hour's wasted time—I arrive at an extremely faint ranch road and a carsonite stake with an AZT sticker on it, the only evidence of actually being on the AZT for days. The faint road leads to a less faint road with no markers of course and I just chance it once again, going right hoping it will lead me to the next supply town, Superior. I luck out and it does after nearly twenty miles of wondering while wandering, and one more night's camp in the where in the hell am I zone.

So now it is decision time. I am way behind schedule and I am supposed to be meeting people at the Grand Canyon for a long awaited journey there. The Grand Canyon trumps all. It is undeniably scenic. Although what I have come through is also scenic, it simply doesn't compare to the Big Ditch. I agonize over continuing for a day in Superior and decide to take a trip to the world famous Boyce Thompson arboretum, to learn the names of the sticker bushes that had been tormenting me. Now I could direct my obscenities more efficiently and with clarity. For example, instead of screaming out "———— I HATE YOU ALL!" to the array of stickies clinging to me I can now say, "Screw you, white bush plant from hell and all the ancestors you supposedly evolved from!" And so on.

I finally decide to abandon ship and bum a ride to Mesa AZ, from an old prospector stopped at a gas station with his old truck. He had to visit his desert shack first where relics from the 1800's abounded. He gave me a couple of large Apache tears—large globs of obsidian—and a small sample of rich gold ore because I showed interest in his stuff. Of course, as required by his profession, he went on and on about how the new world order and the fascist county ordinances had screwed him over big time. I just nodded in support once in a while in exchange for the ride. Apparently he was a real thorn in the county's side, refusing to follow their "capricious and arbitrary" laws and pestering them all the time at county meetings. So I gave him the trail maintenance boss's phone number, saying he was a government agent responsible for implementing the government's land-use plans—and that the guy had mentioned that he hated prospectors because they worked hard and he didn't, making him look bad. No, not really! Basically, the end.

JIFFY POP MADNESS

Once upon a time, a group of us were doing a nine-day trip across the Olympic Mountains. On the evening of the 6th day, we had a campfire, and a Jiffy Pop was produced by a member of the group to cook over it as an after-dinner treat. One could hear the oil sizzling away as he shook it over the flames. After about three minutes, it suddenly burst into flames like it was some sort of Molotov cocktail, causing those in the immediate vicinity to flee. After about ten minutes, the conflagration finally died down, and I went to retrieve the remains, hoping to find a few scorched kernels to eat. The entire thing was consumed except for one tiny corner of the cardboard cover which clearly, and mockingly, proclaimed, "DO NOT COOK OVER AN OPEN CAMPFIRE!"

The next night, yet another Jiffy Pop was miraculously produced by a different member of the group, and the proud owner was eager to show that he was a quick learner. No campfire cooking for him! Instead, he broke out his old Whisperlite stove, lit it off, and proceeded to cook that sucker up.

In order to keep it from scorching, the Jiffy Pop had to be continuously shaken on the open flame. Shake, shake, shake, shake, shake, shake. Once again, one could hear the oil sizzling, and the wafting aroma of popcorn made the anticipation run high. Shake, shake, shake. But suddenly the Whisperlite's pointed, pot-holding tines finally rubbed through the Jiffy Pop's thin aluminum foil, and hot, boiling oil was shaken out to coat the entire stove. The resulting conflagration pretty-much matched the previous evening's excitement, entirely engulfing the stove and its fuel bottle. The whole mess was quickly kicked out of camping range in case it exploded. It didn't, but the Whisperlite's plastic pump was melted beyond usefulness.

The final night of our outing came, and a third Jiffy Pop was really, really miraculously produced by yours truly. So what did I do with the dang thing? I chickened out. I broke open the Jiffy Pop, poured the contents into a large pan, and proceeded to cook it like ordinary popcorn. Sure, the Jiffy Pop container had turned out to be useless deadweight for the entire trip (l could have just brought popcorn and a little cooking oil), but at least I wasn't mocked by a vengeful campfire, and I still had a functioning stove. AND there was a pretty good chance that with my strategy we would actually get to eat some popcorn! Although I was at first scorned for "cheating," it all ended with a chorus of "Ahhs" when I popped off the lid.

BEAR STORIES

YOSEMITE

I was hiking the Pacific Crest Trail setting up camp somewhere in Yosemite when I happened to espy a BIG cinnamon brown black bear that had sneaked up from behind, apparently on tippy toes. He was now closer to my food cache than I was, it being still on the ground. Now if you know anything at all about bears—and I have encountered many, many dozens of them on various hikes—the bold ones are bullies, muggers if you will, as opposed to thieves. If you give them an inch they will take a mile. They are very possessive and to try to take food away from a bear would be a very foolish act indeed. When they see food whose possession is dubious, they will go for it, including food hung from a tree. Possession is everything to a bear, and it's finders keepers if they get their paws on it first. They don't view it as yours unless you are willing to defend it.

Knowing this, my reaction was instantaneous lest all was lost provision-wise. I bluffed him first, before he had a chance to bluff me and literally ran at him screaming that's MY food you sob. It gave him pause and that was the only opening I needed. Avoiding direct eye contact, I began to talk to him in an extremely demanding tone. I picked up a large fallen tree branch and yelled at him as I swung and dashed it to bits against a tree trunk, large chunks flying everywhere, while screaming and that's gonna be your friggin' nose you —— if you go for my food!! Now he was startled and began to back down. I picked up a very large rock and tossed it in his general direction taking great care not to actually hit him. "Yep I can do that to you too, you sob!" He grudgingly walked away and I made the mistake of tossing one last rock that landed in the creek as he crossed it splashing, him big time. He stopped and gave me a real dirty look as if to say "I was leaving and you didn't have to do that." And for a second I thought my bluff was about to fail, but he eventually sauntered on. I slept like a baby that night, my food next to me in my tent because I knew beyond a doubt that he had been so dominated he would not be back. I kept it in my tent in case other bears visited that night, I simply didn't trust hanging it.

The next morning I met a couple on the trail who had camped a quarter mile from me. They said this cinnamon brown bear, obviously the same one I had encountered, rare as they are, had entered their camp and got their hung food despite their pot banging protestations.

From what I have observed, bears consider all food not in anyone's immediate possession to be the spoils of hunting and gathering. Bears will bluff you sometimes to test your resolve to keep the food that is in your possession. It is a bluff because the bear assumes at first that you will vigorously defend your food if someone tries to take it away, just as they would do with their food.

But sometimes a bluff works to pry you away from your food and send you running. Take the case of Bouncing Betty on the Appalachian Trail. Her trick was to saunter down the trail until she ran into some hikers. She would then start jumping straight up and down, in a sort of bounce on all four legs. Some hikers were so unnerved by the sight they took off their packs so they could run away as fast as they could. She then helped herself to the food that was "abandoned" for some odd reason for her enjoyment.

OLYMPICS

The leader of the climbing class knew how to make a little money on the side. He offered to take a group of 11 graduating students of his class on a climb of Mount Olympus. The fee was reasonable, fifty dollars, I think, all permit and tax free, no doubt! The approach we took went up the Quinault River to Low Divide, then down to

Elwha River and up the Elwha to its headwaters. The route then traversed across Bear Pass, the exit point for the Bailey Range traverse. It then entered the upper Queets River basin, headed up the Humes Glacier, across the Hoh glacier, up all three summits of Olympus and then down the Blue Glacier to the Hoh River for a sixteen-mile hike out. As we went up the Elwha River we finally reached the end of the trail and started using game tracks instead. One of these tracks led through a large thicket of brush. Simply because of the morning dew, the brushes' branches were so weighed down that they drooped to the ground. Such is hiking the Olympics! You can get soaked without a drop of rain falling. I volunteered to be first through wet stuff. I put on all my rain gear and proceeded on through, shaking the branches as I went to relieve them from their dewy load. Others then following me would stay relatively dry without the need for sweat producing rain gear.

After perhaps one hundred yards, I burst through the brush into a clearing meadow, and gulp, there she was, a rather large black bear with her tiny cub. The eyes of the cub had not fully opened yet, and apparently all it could see was shapes. Instead of running toward its mother, it ran straight over to me, got behind me and nudged against my leg. I could feel it shivering in fright. I could also see the look on mama bear's face, a look that seemed to suggest that she wasn't at all happy about the situation. And neither was I. I started shivering in fright too, as I stumbled backward while yelling,

IT'S A BEAR!

The squinting cub then caught on that I wasn't one, and ran toward and past mom into the forest. Happily, she followed it. And only the person directly behind got to see the fleeing bear.

On one week long high alpine traverse in the Olympics in late August, we counted twenty-five separate bears. Most were seen in high valleys as we crossed above them on a ridge line. It seemed every high valley on either side of the ridge had its own bear slowly grazing uphill, following the ripening mountain blueberries and huckleberries. So it seemed nothing out of the ordinary as we crossed a meadowed hillside to see one peacefully grazing perhaps one hundred yards below us. But as the two lead hikers got near the middle of the very steep hillside meadow, the bear suddenly reared up and peered intently at them. Then quick as a flash it bolted uphill, its muscles rippling as it went. The bear covered the one hundred yards in about five seconds. It got within ten feet of the traversing hikers when it stopped, reared up, sniffed the air, and just as suddenly bolted away back downhill before slanting off into the forest. We figured it wasn't a bluff gone bad, but a case of mistaken identity. Perhaps the bear, with its poor eyesight, simply saw a moving shape uphill that it thought was another bear, entering his territory to steal his forage. When he got close enough, his nose kicked in, and the odor from the sweating stinky hikers was so overwhelming he just had to run off into the woods to puke. Or maybe he was just afraid of humans and his nose didn't pick us up because we were downwind from him. I asked the two what they were thinking as they watched the bear charge them and they said they just had enough time to say "What are we gonna do?" and by then the bear was upon them.

We camped near a snowmelt tarn at about 4,500 feet. The summer sun had warmed the small, quarter-size pond into the seventy-degree range, so I decided to take a bath. I noticed what appeared to be submerged elk tracks near the shore as I entered. After bathing and redressing I started back toward camp when I spotted a bear approaching the tarn from a different direction. I stopped to watch him. He proceeded to wade right into the tarn, submerging himself up to his nose in an apparent attempt to get rid of the parasitic bugs that plagued him. After about fifteen minutes, he got out and shook off like a wet dog. I realized, given all the submerged tracks, that this was probably a daily event, and given a bear's bathroom etiquette and odor and knowing beyond a doubt that he had not soaped up and showered before entering the tiny pool, suddenly I didn't feel so clean.

YELLOWSTONE

When I worked in Yellowstone National Park in the summer of 1972, there was a man killed very near Old Faithful by a large grizzly bear. He was camping in an illegal area and had gone to the Old Faithful area to buy supplies, leaving a friend to guard their camp. His friend was cooking dinner while he was gone, and a Grizzly bear and her cub entered the camp to see what the odors were all about. Eventually the sow entered their fairly large family style tent to search for food. Just then the guy returned and was warned that there was a large grizzly rummaging around in the tent. Well this guy must have thought he was in Jellystone National Park, and that was Yogi Bear and Boo Boo stealing his stuff. He proceeded to grab a large frying pan, entered the tent and swatted the bear in the butt with it. Now I have read the grizzlies are pretty much like buffalo and moose in that they are ALWAYS in a bad mood. So I imagine getting swatted on the butt didn't help much to improve her demeanor. According to his friend, the bear immediately turned around and swatted HIM in the head with a massive paw, killing the guy instantly, and proving yet again that Darwin was right.

His friend didn't hang around to become bear seconds. He went running to report the incident to the rangers. When they got back, the bears were gone, along with part of his friend's body. I guess she was pretty hungry. They eventually tracked down and shot the grizzly, assuming she now had a taste for humans. They knew they got the right bear because human hair was found in her stomach.

On a happier note, we worked out in the boonies as surveyors in the Park. One time I got to witness how powerful bears can be. One of the rude kind ambled by us one day—a big black bear with a tongue lapping just like a dog who refused to acknowledge our presence. It was like we were being shunned. I've had other bears do the same thing in the Olympic Mountains. He proceeded to go over to a large Lodgepole pine stump, placed a front paw on it and literally ripped it right out of the ground, his shoulder doing most of the work in rippling bursts. It would have taken me all day with a pick axe and saw to get that stump out of the ground. And this bear did it almost as efficiently as a stick of dynamite or two. He then enjoyed the bugs and larvae under it.

Okay this isn't a bear story, but a buffalo story instead. I was at West Thumb concession area and decided to take a tour of the geysers near there. As I walked along I noticed a buffalo ahead with a bunch of tourists taking pictures of it. One guy was particularly aggressive, getting closer and closer to the buffalo. Now they WARN you not to do that, and this guy was about to find out why. He literally got within two feet of the 1,500-pound buffalo's head with his camera when that beast suddenly decided he had had just about enough of this rude twit. So he butted him. The guy started running away after stumbling backward and the buffalo followed him. The amazing thing was they both ran past dozens of tourists, any one of which could easily have been gored by the charging buffalo. But oh no the buffalo had it in for this one guy. He chased him several hundred feet into a parking lot, and they played ring around the car for a while before the buffalo finally calmed down and left. Taught him to show a little respect!

THE GOATS OF OLYMPC NATIONAL PARK SAGA

Back in the 1980s, I remember reading a headline in our local newspaper that stated the mountain goats were not native to the Olympic Mountains and that the park officials were planning to remove them. So began a decade long theater of the absurd. The park had to prove three things in order to justify removing such a high order of mammals from a Park it was sworn to protect. One was that the goats were not native. Two, they could not be controlled by any other means than total removal. And three, they were doing irreparable damage to the alpine ecosystem of the Olympic mountain range.

Now I have some understanding of scientific method, having taken college lab course in both biology and chemistry. So as I read the way the studies were conducted to prove the goats were bad for the Olympics, red flags went up. It just seemed to me that there was a lot of bias in the studies and that a conclusion had been reached—the goats were doing harm—and that evidence had been twisted to support that conclusion.

Pictures of areas showing terrain both before and after goats congregated in an area were shown to prove that harm. It just seemed too simple and unnatural to me to deliberately congregate goats around a salt lick, let the salt leach into the soil and then to say this is the fate of Olympic National Park if we don't eliminate the goats—a large horse pen denuded of vegetation. So when I read that the Park service was holding public meetings on the issue, I decided to go. It soon became apparent that the meetings were to inform the public of a policy already decided and that the public "input" meetings were just a sort of lip service. I never felt so patronized in my life! The attitude was, the learned scientists and you are the ignorant, emotional public whose opinions are in our goat policy implementation's way. In short, it was a show for presumed sixth grade intellects.

This went on for several years with the Park Service gaining the support of such organizations as the Sierra Club and the Mountaineers. The National Forest Service that abuts the Park refused to remove their goats as requested to so they would not wander into the Park and damage it. They stonewalled saying they thought the goats were native. Finally the decision was made to move ahead with the goat removal plan and to shoot old Billy right between the eyes, as one member of the outdoor club I belonged to stated. I was considered a bleeding heart to some and basically told to take things unemotionally, and just let the NPS do its job.

The Park service then decided to play tricks with statistics. In order to placate the public outcry somewhat, they agreed to remove only half of the goats. But they inflated the number of goats in the Olympic mountains. They claimed one thousand were there instead of the actual five hundred. So removing half of the fake one thousand was in actuality removing the entire goat population of five hundred. They were caught in this lie by me at a public meeting and had no valid explanation, and instead quickly changed the subject by saying errors could easily be made in a goat population survey.

But alas for the Park Service, somebody opposed their plan that had just as much money for lawyers as the Park service did and a representative there heard what I had to say about spinning statistics. The Fund for the Animals put a juggernaut into the plan by getting a long time local politician to step in and give a stay of execution to the goats pending an independent scientific review of the park's conclusions—that the goats were not native and doing irreparable harm to the Alpine Ecosystem of Olympic National Park.

Over one year later the findings of the independent scientific study came to some conclusions. When the Park read it they rushed to put a headline on their website stating that they were vindicated! The study agrees the goats were not native! So with a heavy heart, I downloaded the independent scientific review and prepared for more chortling from my peers. But as I read the study it dawned on me that the Park was actually being condemned by the study, albeit with soothing language.

As to the native issue, the study said that they could not disprove the contention by the park that the goats were not native, but they noted that does not necessarily mean they were correct in the non-native status contention. They said imagine the outcry if sometime in the future, evidence was produced to show the goats to be native after all. Therefore the conclusion was that goat removal should not be decided on the native or nonnative issue because neither could be proven beyond a reasonable doubt.

The independent study also stated that the contention that the goats were doing irreparable damage to the alpine ecosystem of Olympic National Park to be "unnecessary alarmism." It further stated that the studies done by the Park Service to prove that "unnecessary alarmism" was, just as I had first suspected, "bad science at its worst"

And just to put the icing on the cake, one scientist stated that the mountain goats were actually doing more good for those rare plants the park was worried about getting munched or trampled to death, than bad. She noted that just because a plant is eaten or trampled, doesn't mean that it won't come back from the roots. She also noted that the sleeping wallows the goats made allowed for rare plant species to get a hold in the otherwise impenetrable mountain heather, so removing the goats would do more harm than good to them.

THAT was the end of that. Not so much as a peep from the Park Service about mountain goat removal since!

In retrospect it is easy to laugh at the Park's goat removal policy logic. After all, Goat Rocks wilderness is only fifty to seventy-five air miles from the Olympics, and there are ten times the number of goats there than in the Olympics in a far, far smaller area—and not a peep about them damaging the nearly identical alpine ecosystem that exists there.

Yeah, I rubbed it into those who insisted the goats be removed, what can I say. Goat killers!

REVENGE OF THE GOATS!

During this era, I did a cross country trip deep into the core of the Olympics and ended up in little-used, trailless Muncaster Basin. Our group of five was on a peak bagging expedition, and Muncaster was on our list. On the way up we espied several mountain goats and my thoughts were on them as we reached the summit. We soon might not be seeing any more of one of five large mammal species within the Park.

On the summit was a shed with antennae sticking out of it. Out of curiosity I walked over to it to have a look. Several access panels were padlocked but then I noticed one with an unsecure lock. I took the lock off and removed the access panel. Inside was what appeared to be a large battery, and a terminal board with switches for about ten backcountry ranger stations and Park Headquarters. It appeared to be (and I later confirmed) a relay panel to amplify a relatively weak signal from a backcountry walkie-talkie and send it on to all other backcountry ranger stations and park headquarters to obtain help in case of emergencies. A walkie-talkie labeled Muncaster was wired into the terminal board and its switch was in the on position. I suddenly realized that if I pushed the walkie-talkies button, I could talk to all the backcountry rangers and Park headquarters simultaneously. I pushed the button and could hear static. I had an idea and since the statute of limitations has expired I can admit to bringing that idea to fruition. I pushed the button again and blurted out, "ATTENTION, ATTENTION this is the goats speaking! Up yours, Maureen! [ONP Superintendent.] Oh, look, here's some more of those endangered plants. CHOMP, CHOMP, yummy, yummy. They're my favorites. Come up and make us stop, Maureen, baaaa, baaaaa!"

The three guys with me on the mountain stared at me in disbelief, their jaws dropped to their chests. They then laughed and shook their heads.

Several minutes later one of them said, "Look, a helicopter is headed this way!"

For a brief moment I was dumbfounded that such a quick response could be mustered to my obscene message. They flew quite close but continued on. It became apparent that they were checking on a smoldering logging slash cut fire on the border of ONP, and not my miscreant deed. They flew back in a different direction, much to my relief. I suppose obscene phone calls from persecuted goats was low on their priority list. I always wondered if Maureen was in her Port Angeles office at the time of the goat call. I certainly hoped so. She was soon transferred to a Washington, DC, monument because of the public relations debacle.

Tamarisk and Tamarack
The journey there, the journey back
One's up high, one's down low
At sixty-two am torn, which way to go
Mystic meadows, with my pack
A heavy weight upon my back.

Or Grand Canyons deep and low,
And just go with, the river's flow

GOLD LEGEND MEETS GOOGLE EARTH

Once I was thinking about doing some gold panning just for outdoor fun, so I Googled places to go in Washington State. Turns out there were few and in between, most near Blewett Pass. I happened upon the Ingalls Gold Legend and became intrigued so I read up on it figuring it was mostly fantasy like many gold legends are.

The story basically went like this. A military scouting party led by Captain Ingalls in 1861 was scouting out an area near Mount Stuart looking for land suitable for farming by colonists. Ingalls somehow got separated and went the wrong way. Instead of going up Icicle creek where the main party went, he eventually ended up in Enchantment Lakes, apparently by following what is now known as the Snow Creek trail. There he saw three lakes connected by creek descending into Ingalls creek. He found the color of the middle one to be interesting and descended the drainage to check it out. On the shore he found a lot of gold, many tons of it mixed with quartz just coming out of some sort of mother lode. Instead of going back the way he came, he descended to the creek now named for him—Ingalls—and eventually got back to his fort near Wenatchee to rejoin his men. He told no one. After getting out of the Army he settled in Oregon and then made a trip in the late 1860s with his son and several friends to find his gold.

While traveling through thick forest, a branch pushed forward by Ingalls' pick swung back and hit the guy behind him hitting the trigger of a rifle. It went off, and Ingalls was hit in the back. He died two days later after giving a precise description of where to find the gold. They abandoned the search and sadly went back to Oregon. Later his partner moved to and homesteaded near the mouth of Ingalls creek. They spent the next twenty years searching for the gold, and never found it. Others learned of the legend and the Enchantments were picked over with a fine toothed comb for gold. Nothing.

So what happened to the gold?

Well first some clues. Ingalls almost certainly would have ascended into the Enchantments via easy Snow Creek. It was right along the way the military party was going along Icicle Creek, and was easy enough for a horse to do They stayed straight and he probably went left, eventually topping out in what is now known as the Enchantment Lakes Wilderness area, rife with alpine lakes.

The key to the story is that he was with a horse. Now the story says he descended from The Enchantments to Ingalls Creek. You can't just go anywhere through the ridge above Ingalls Creek to do that, 95 percent of it is class three climbing and above and passage would seemingly be impossible for a horse.

So I got on Google Earth and asked myself, where could a horse possibly descend from way up there down to Ingalls Creek, where no human trail dared to go? As I scanned the terrain with the altitude tool, it became quite apparent that there was ONLY one drainage that a horse could possibly go down, and it was very close to where Snow Creek plateaued out, in fact the very first drainage to its left, Crystal Creek. But Google Earth showed no lakes on Crystal Creek. I scanned the drainage and about five hundred feet down there was a long flat spot where lakes could exist,

Even more astonishing was there was an obvious fault line in the cliffs above the flat spot, a fold in the Earth quite visible. Mount Stuart, just like Yosemite, is basically one giant piece of granite—a batholith—eroding away. There is very little gold or anything else in the block—but along its edges there can be, such as the case in Yosemite where plenty of gold in intrusive quartz veins was found on the northern edge of the batholith. Although some gold has been found in Ingalls creek, nothing like in California.

So this fault line I was looking at seemed like a plausible place for a mother lode to me, being on the edge of the Mount Stuart batholith. So where did the lake and gold go?

There was one more part of the legend, a final clue. A very significant event occurred between the time Ingalls found his gold and when his partner went to look for it a few years after Ingalls died. The biggest earthquake ever recorded in Washington, which happened in 1872. So the presumption was that landslides caused by the earthquake buried everything.

Well there was only one way to find out, go there. I backpacked up Ingalls to Crystal Creek and then day packed up the steep creek's drainage to see what I would find. When I got to the flat spot half a mile up in elevation I could see that the last part of the legend had been true. The entire area was covered twenty feet deep in rock debris from an enormous slide, no way on Earth to get to the bottom of it to see if there was any gold there. The way forward to the hillside above the debris slide was impassable, many, many ten-foot drops between supersized boulders blocked the way up to where I thought the fault line was.

I'm certain the gold is there, the legend seems based in facts, the geology supports the existence of a mother lode in the area and there is simply no other place it can be but on the upper reaches of Crystal Creek, thanks to the horse and Google Earth clues. One more thing—it's called Crystal Creek I presume because crystals were found in it? That would imply quartz, a rarity in the granite of the area, and quartz infers the possibility of gold.

Due to commitments at home, I was pressed for time so I gave up and went out in the ninety-five-degree afternoon heat. Something rather bizarre happened on the way out. I stopped at a camp spot near a cliff to take a long water break. There was a sort of minor cave in the cliff and I went to check it out. There was frigid air coming out of a cavity, so cold you could easily see your break, I'd guess around thirty-five degrees Fahrenheit—twenty feet away, it was ninety-five degrees Fahrenheit. What a find on a hot day! Must have been a big crack or fault line in there somewhere. Then I left, my dreams of hauling forty pounds of gold out on my back shattered.

If I were to do it again I would take the two day trip up to the Enchanted Lakes via Snow Lake the way Ingalls must have gone, camp at the top of the Crystal Creek drainage, a hop, skip, and a jump from the top of Snow Creek, making it a simple day hike down to the slide area—maybe 750 feet down an easy slope—where you could explore the area extensively for many hours in the area of the fault line above the slide debris—even one lousy fist-size nugget would be worth the effort, right?

MY RED WING HIKING BOOTS

When I signed up for the Olympic College of Bremerton's basic mountaineering class, I was told I should get expensive mountaineering boots for the course, but it was optional. I decided to use the Red Wing hiking boots I already had for backpacking, for the climbs.

I couldn't help but notice that a lot of the people using mountaineering boots in the class including the assistants suffered bad blisters because they weren't really made for hiking in on the approach, more for use on hard crusted snow because their stiffness made it easier to kick a step. Great on Rainier I suppose but horrible on any other nonsnowy terrain.

I wasn't getting any blisters and they worked just fine in most snow conditions. If it became icy and difficult to kick steps, I put crampons on them.

When I started to long distance hiking, I was told I would need lightweight glorified tennis shoes—I couldn't help notice that the people who used them were getting terrible blisters and that their shoes wore out rather quickly. So decided to use the Red Wings on the PCT and AZT and once again never got a blister. They lasted for around two thousand miles a pair.

When I did Brewer's Buttress with the Canadian guides by pure chance, I was wearing Red Wings because I had no rock climbing shoes. I now know they are good up to 5.8!

Rainier, Grand Canyon, Grand Teton, 5.8, PCT, Red wings did fine on them all—what could possibly be more versatile than that for a boot?

I'm not saying Red Wings are the only boot you should use, just that good hiking boots are very versatile. I have very wide

EE feet, almost like a duck, and needed a brand that provided that. Tried Red Wings once and the results were so good I never switched boots again. I decided I didn't need to fix what was not broken.

THE TWENTY-POUND PACK

I've had many people comment on how light my pack was when comparing it to theirs, asking how I do it. After thirty years of backpacking, I hope that, by keeping an open mind, I have learned a thing or two about knowing what to throw away and knowing what to keep. It became obvious to me that the old Boy Scout adage of being prepared for any eventuality was the root cause of sixty- to eighty-pound packs. It also was obvious that trudging up three thousand feet with such a load took away a lot of the enjoyment of a backpack trip. In addition, by the time I got to camp I was so exhausted and riddled with ailments such as blisters, sway back, and symptoms similar to mad cow disease that the only thing I wanted to do was lay around in the tent until 10:00 a.m., followed by a knee busting descent back to the car. I decided to do something about it rather than to continue to suffer until all I wanted out of being outdoors was three-mile day hikes up some ubiquitous flat river bottom trail.

So I started to question the Boy Scout credo and started to throw out things like axes, shovels, eight-pound tents, six-pound ropes, crampons just in case, five-pound Coleman sleeping bags rated to negative twenty degrees Fahrenheit, leg splints, first-aid kits adequate for nuclear war, Coleman stoves and lanterns, short wave radios, Eskimo jackets, porta potty, one clean set of clothes per day, and a Bowie knife And somehow it began to dawn on me that carrying three quarts of water in the Olympics was a bit much when there was a water source every hundred yards.

Most backpacking weight is in the Big Four: shelter, sleeping system, clothing and pack. Let's start with the pack. Yeah seven-pound Gregorys are probably great for carrying sixty-pound loads. But thirty-five pounds? There are plenty of backpacks out there around four pounds more than capable of carrying thirty-five pounds of total weight that includes water and food for a weeklong trip. All that is necessary is a good fit, meaning the shoulder straps should be on your shoulders, with no air between your shoulders and your straps.

There are several sub two-pound down bags around rated down to twenty degrees, more than adequate for three season hiking. "But down is useless when wet," is the oh-so-common refrain of those who prefer synthetic bags. My response is that I have backpacked for thirty years, spending many hundreds of nights in a down bags that have NEVER been wet (well, except for washing) Do you really think this is blind luck considering all the major moisture we get around here? With a minimum of creative effort, a down bag can be kept dry in the worst of downpours (hint: think plastic bags).

There are several solo tents around that are three pounds. If you insist on tenting with someone, enjoying their body and foot odor, gas, and snoring (usually in unison), then you can share a roomier six-pound tent. Sweet dreams!

That leaves clothing. Not counting what I am wearing, the clothes in my pack weigh less than four pounds: Jacket 14 ozs, rain/wind parka 12 ozs, Polartec 100 pants and shirt 16 ozs, gloves 3 ozs hat 2 ozs, and extra socks 7 ozs. Throw in a towel and a couple of wash cloths (one for dishes) and the clothing bag is around 3.5 pounds. I wear a Supplex nylon shirt and pants that can be washed daily and hung to dry overnight while I wear the Polartec stuff. I found that all other clothing was really unnecessary and just dead weight. Some people take three or four suitcases of clothing when traveling by air. Others just a carry-on bag. Like Thoreau said, simplify, simplify.

Here is a summary of what I took on my six month PCT hike. All weights are in ounces

PACK	67 (with 4 removable pockets)
SLEEPING BAG	28
PAD	11 (therma rests are WAY too heavy and useless if punctured; try finding a cactus hole)
TENT	48 (with stakes)

BIVY SACK	17
2 COMPACTOR TRASH BAGS	3 (pack fly and clothing/sleeping bag bags)
KNIT HAT	2
POLARTEC 100 SHIRT	8
POLARTEC 200 JACKET	13
PARKA	12
POLARTEC 100 PANTS	
EXTRA SOCKS	5
EXTRA LINER SOCKS	2
GLOVES	3
2 WASH RAGS, 1 TOWEL	2 (one for dishes)
SIERRA ZIP WOOD STOVE	17
2 POTS AND LID	9 (including grip handle)
FORK, SPOON, 2 CUPS	4
WATER CONTAINERS	5 (I like dromedary bags)
CAN OPENER	1/4 oz
TOILET PAPER	6 oz (enough for a 9-day trip)
TOILETRIES	4 (comb, toothpaste and brush, floss, soap)
FIRST AID	6 (various and sundry pills, Band-Aids, compress, whistle, elastic bandage, chopsticks)
INSECT REPELLENT	2 (very little needed if you wear pants and a long-sleeved shirt.)
KNIFE TOOL	3
2 CIGARETTE LIGHTERS	1
MAP AND COMPASS	2 (use back of map for diary)
REPAIR KIT 2 (tape, needle, thread)	
STRING	3 (hang laundry and food)
2 Mini LED lights	1
4 EXTRA AA BATTERIES (for stove and camera)	3
ALTIMETER WATCH	wear—do not count

PENCIL	1/4 oz
SUNCREAM	2 (for snow travel only; a tan, pants, and a long-sleeved shirt otherwise)
SUNGLASSES	1 (only used for snow travel)
DIGITAL CAMERA	6
ICE AX	14 (long, not short)
TOTAL	310.5 = 19 lbs 6.5 ozs

I double-checked the weight by putting together the entire pack, and it actually weighed slightly less at nineteen pounds three ounces. A desert trip would be lighter.

There are plenty of backpackers around today who would consider my pack to be overly heavy. Many people get their base weights down to less than fifteen pounds. They opt for tarps instead of tents, alcohol stoves that weigh less than an ounce, thirty-five-degree bags, packs with no waist belts that weigh less than two pounds and lightweight water carriers of the Platypus type. They toss the ice ax.

I try to take a more conservative approach, following my own idiosyncrasies such as taking a bivvy sack for those REALLY wet or cold nights and more camping options, and an ice ax in case I want to go climbing. I like a wood-burning stove because I don't have to carry or ration fuel. I like a three quart bomb-proof dromedary bag because I can use it for a hot shower. So the next time you are struggling up some god-awful steep hill with a sixty-pound load, contemplating switching forever to those cookie cutter, flat, river-bottom day hikes suitable for disillusioned boy scouts, the gravity challenged among us, heavy smokers and those who just plain abhor putting their antiperspirant to the test, remember: it doesn't have to be that way, so lighten up!

YOSEMITE MYSTERY ANSWER

After they got down, they took a break. While doing, so a bear came along, snatched one of their packs, and ran off into the woods with it. He obviously wasn't interested in anything in it but food, so there it lay until you found it.

ABOUT THE AUTHOR

David Cossa, aka Mountain Dave, was raised in Colorado. As a boy, he became interested in hiking by taking the JFK challenge to hike fifty miles in twenty-four hours. He became interested in mountaineering thanks to a church camp he attended that climbed a minor peak as one of their outings.

After a four-year stint in the Navy as an aircraft electrician, he got a job a working as surveyor in Yellowstone, and eventually, he moved to Port Orchard, Washington, to work as an electrician at Puget Sound Naval Shipyard. He took the Olympic College of Bremerton's Basic and Intermediate Mountaineering classes and, after a rocky start, became a climb leader for them for a number of years, leading numerous Rainier climbs for the students as a graduation present. He has been many times to both the Canadian Rockies and Grand Tetons where he summited major peaks, a number of them solo, and eventually bagged 250 peaks throughout the western US and Canada.

He hiked the 2,650-mile Pacific Crest Trail in one season, the desert part of the Arizona Trail, and in the Grand Canyon when his interest went to long-distance hiking. Currently he is hoping to finish summiting the highest peaks in the Lower 48, having already climbed all the western ones, and to someday finish the AZT.

CPSIA information can be obtained
at www.ICGtesting.com
Printed in the USA
BVHW012134291119
565228BV00006B/112/P